Selected Titles in This Series

(*Continued in the back of this publication*)

Black Box Classical Groups

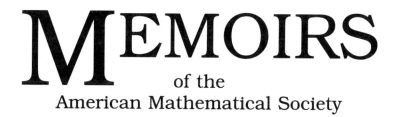

MEMOIRS
of the
American Mathematical Society

Number 708

Black Box Classical Groups

William M. Kantor
Ákos Seress

January 2001 • Volume 149 • Number 708 (third of 4 numbers) • ISSN 0065-9266

American Mathematical Society
Providence, Rhode Island

1991 *Mathematics Subject Classification.*
Primary 20B40, 20G40; Secondary 20P05, 68Q25.

Library of Congress Cataloging-in-Publication Data

Kantor, W. M. (William M.), 1944–

    Black box classical groups / William M. Kantor, Ákos Seress.

      p. cm. — (Memoirs of the American Mathematical Society, ISSN 0065-9266 ; no. 708)

    Includes bibliographical references.

    ISBN 0-8218-2619-0 (alk. paper)

    1. Permutation groups. 2. Matrix groups. 3. Algorithms. I. Seress, Ákos, 1958– II. Title.

III. Series.

QA3 .A57 no. 708

[QA175]

510 s—dc21

[512′.2]   00-046914

Memoirs of the American Mathematical Society

This journal is devoted entirely to research in pure and applied mathematics.

Subscription information. The 2001 subscription begins with volume 149 and consists of six mailings, each containing one or more numbers. Subscription prices for 2001 are $494 list, $395 institutional member. A late charge of 10% of the subscription price will be imposed on orders received from nonmembers after January 1 of the subscription year. Subscribers outside the United States and India must pay a postage surcharge of $31; subscribers in India must pay a postage surcharge of $43. Expedited delivery to destinations in North America $35; elsewhere $130. Each number may be ordered separately; *please specify number* when ordering an individual number. For prices and titles of recently released numbers, see the New Publications sections of the *Notices of the American Mathematical Society*.

Back number information. For back issues see the *AMS Catalog of Publications*.

Subscriptions and orders should be addressed to the American Mathematical Society, P. O. Box 845904, Boston, MA 02284-5904. *All orders must be accompanied by payment.* Other correspondence should be addressed to Box 6248, Providence, RI 02940-6248.

Memoirs of the American Mathematical Society is published bimonthly (each volume consisting usually of more than one number) by the American Mathematical Society at 201 Charles Street, Providence, RI 02904-2294. Periodicals postage paid at Providence, RI. Postmaster: Send address changes to Memoirs, American Mathematical Society, P. O. Box 6248, Providence, RI 02940-6248.

Contents

Abstract

If a black box simple group is known to be isomorphic to a classical group over a field of known characteristic, a Las Vegas algorithm is used to produce an explicit isomorphism. The proof relies on the geometry of the classical groups rather than on difficult group-theoretic background. This algorithm has applications to matrix group questions and to nearly linear time algorithms for permutation groups. In particular, we upgrade all known nearly linear time Monte Carlo permutation group algorithms to nearly linear Las Vegas algorithms when the input group has no composition factor isomorphic to an exceptional group of Lie type or a 3–dimensional unitary group.

Key words and phrases: computational group theory, black box groups, classical groups, matrix group recognition

1991 Mathematics Subject Classification
Primary: 20B40, 20G40
Secondary: 20P05, 68Q25, 68Q40

This research was supported in part by the National Science Foundation and the Alexander von Humboldt Foundation.

Received by the editor April 15, 1997.

1 Introduction

In a number of algorithmic settings it is essential to take a group known to be simple and produce an explicit isomorphism with an explicitly defined simple group such as a group of matrices. At present, the main examples of this need occur in finding Sylow subgroups of permutation groups [Ka3, Ma, KLM, Mo1, Mo3] and in recognition algorithms for quasisimple matrix groups [CLG2, Ce]. Whereas the preceding references deal with special instances of this type of question, a recent breakthrough by Cooperman, Finkelstein and Linton [CFL] studies a much more general setting. By considering the case of $\mathrm{PSL}(d, 2)$ given only as a *black box group*, they provide a framework for handling many such questions simultaneously. In this paper we proceed in a direction suggested by their work, *providing fast explicit isomorphism algorithms for all classical groups over all finite fields.*

The elements of a black box group G are assumed to be coded by 0-1 strings of uniform length N, and G is specified as $G = \langle \mathcal{S} \rangle$ for some set $\mathcal{S}$ of elements of G. Hence $N \geq \log |G|$, so for classical groups of dimension d over $\mathrm{GF}(q)$ we have that $d^2 \log q$ is $O(N)$. Our main result is as follows:

Theorem 1.1 *There is a Las Vegas algorithm which, when given a black box group $G = \langle \mathcal{S} \rangle$ isomorphic to a simple classical group defined on some vector space over a field of given characteristic p, verifies that there is, indeed, an isomorphism, and finds the following:*

(i) *The field size $q = p^e$, as well as the type and the dimension d of G; and*

(ii) *A new set $\mathcal{S}^*$ generating G, a vector space $\mathrm{GF}(q)^d$ and a monomorphism $\lambda \colon G \to \mathrm{PSL}(d, q)$, specified by the image of $\mathcal{S}^*$, such that $G\lambda$ acts projectively on $\mathrm{GF}(q)^d$ as a classical group defined on $\mathrm{GF}(q)^d$.*

Moreover, the data structures underlying (ii) *yield deterministic algorithms for each of the following:*

(iii) *Given $g \in G$, find $g\lambda$ and a straight-line program of length $O(N)$ from $\mathcal{S}^*$ to g;*

(iv) *Given $h \in \mathrm{PGL}(d, q)$, decide whether or not $h \in G\lambda$; and, if it is, find $h\lambda^{-1}$ and a straight-line program of length $O(N)$ from $\mathcal{S}^*$ to $h\lambda^{-1}$; and*

(v) *Find a form on $\mathrm{GF}(q)^d$ involved in the definition of G as a classical group, if $G \not\cong \mathrm{PSL}(d, q)$.*

Finally,

(vi) $|\mathcal{S}^*|$ *is* $O(N)$;

(vii) *The algorithm in* (i,ii) *takes* $O(\xi q N \log N + \mu[q^{3/2} N^2 \log^3 N + N^{5/2}] + |\mathcal{S}|[\mu q^{3/2} N + N^{5/2}])$ *time, where* ξ *is an upper bound on the time requirement per element for the construction of independent,* (*nearly*) *uniformly distributed random elements of* G *and* μ *is an upper bound on the time required for each group operation in* G; *and*

(viii) *The algorithms in* (iii) *and* (iv) *take time* $O(\mu q^{3/2} N + N^{5/2})$ *and* $O(\mu N)$, *respectively.*

While the main part of the theorem is (ii), the other parts are designed to make the result effective: all elements of G and $G\lambda$ are made accessible for further computations. For example, (iii) produces the image $g\lambda$ by writing a straight-line program from $\mathcal{S}^*$ to g and then evaluating the resulting straight-line program in $G\lambda$ starting from $\mathcal{S}^*\lambda$ (cf. **2.2.5**); and a similar statement holds for (iv). Keeping track of the steps used to construct $\mathcal{S}^*$ from $\mathcal{S}$, we also *obtain straight-line programs from the* original *generating set* $\mathcal{S}$ *to any given element of* G (compare [BSz]). Note that there is some ambiguity about d in some cases, where we use the smallest d.

When q is polynomial in N ("tiny" in computer science terminology) our algorithms run in polynomial time. For example, this is the case in the permutation group setting (cf. Section 8). Whereas the theorem may give the impression that complexity is our only goal, we are also concerned with its practical uses. The second author has implemented the $\mathrm{PSL}(d, q)$ and $\mathrm{PSp}(2m, q)$ cases of the theorem in the group theory system **GAP** — leading in turn to significant improvements in the algorithms that are incorporated here.

The theorem produces randomized algorithms, and in particular requires a method of selecting nearly uniformly distributed random elements of G; a fundamental algorithm for this is provided in [Ba1] (cf. Theorem 2.2). We recall that a randomized algorithm is called *Las Vegas* if the output is correct when there is an output; the algorithm may also report failure, which occurs with probability less than $1/2$. Higher reliability can be achieved by repeated applications: after $\log(1/\varepsilon)$ runs, the probability that we do not have an output is at most ε. In contrast to Las Vegas algorithms, a *Monte Carlo algorithm* is a randomized algorithm which always returns an output, but it is not guaranteed that the output is correct; an upper bound for the probability of an incorrect answer can be prescribed by the user. Babai [Ba2] provides a thorough discussion of these notions, and of practical *vs.* theoretical notions of efficiency of randomized algorithms for groups.

In Sections 3–6 we will handle the various types of classical groups. It is unfortunate that we have found it necessary to consider the types separately. That certainly has added to the length of our proof. However, the details differ significantly in the various cases, the orthogonal groups being especially intricate. We remark that Section 3, the case of the linear groups $\mathrm{PSL}(d, q)$, is written in more detail than Sections 4–6, where in some cases subroutines that are exact analogues of ones in Section 3 just refer to that section. On the other hand, this case is the least interesting within this paper since it is the one in which transvections behave especially well.

Throughout Sections 3–6 we assume that the type, the size q of the underlying field, and the dimension d of G are all known. In **7.2.1** we will use a polynomial time Monte Carlo algorithm to find (at most) seven possibilities for the triple $(q, d, \text{type of } G)$. If the procedures in Sections 3–6 return an output for one of these triples, correctness is verified in **7.2.2** by using a presentation of G.

The parameter ξ is discussed in **2.2.4**, where we will see that it can be expressed as a polynomial function of N and μ, so the running time of our algorithm is μ times a polynomial in N and q with computable small exponents. It would be highly desirable to have a version of this theorem using $\log q$ in place of q in (vii) and (viii), but such a result seems quite difficult—even for the group $\mathrm{PSL}(2, q)$! Of course, this distinction disappears if q is bounded, as in [CFL].

The standard examples of black box groups are matrix groups and permutation groups. For the former we obtain an immediate

Corollary 1.2 *Given a quasisimple matrix group $G \leq \mathrm{GL}(a, F)$ (for any finite field F) such that $G/\mathrm{Z}(G)$ is isomorphic to a simple classical group over $\mathrm{GF}(q)$, G can be identified constructively by a Las Vegas algorithm, using an isomorphism from $G/\mathrm{Z}(G)$ to a classical group of matrices over $\mathrm{GF}(q)$ modulo scalars. The algorithm runs in time polynomial in a, $\log |F|$, and q.*

This is a *constructive* recognition algorithm, as contrasted to the nonconstructive ones appearing in [CLG1, NP, NiP1, NiP2, BKPS] where only the isomorphism type of G is determined. The first constructive algorithms for simple permutation groups are in [Ka3], followed by [Ma, KP, Mo1, Mo3].

The first constructive algorithm for matrix groups was obtained by Leedham-Green [LG] in 1992 for groups between $\mathrm{SL}(d,q)$ and $\mathrm{GL}(d,q)$, given as acting on $\mathrm{GF}(q)^d$. This has been improved and implemented in [CLG2]; the symplectic case is discussed in [Ce].

One consequence of a constructive algorithm is a *membership test*, a type of algorithm of fundamental importance for dealing with groups algorithmically. In the matrix group context, this allows us to decide whether or not any given element of $\mathrm{GL}(a,F)$ is in G, $\mathrm{N}_{\mathrm{GL}(a,F)}(G)$, or $G \cdot \mathrm{C}_{\mathrm{GL}(a,F)}(G)$. More generally:

Corollary 1.3 *Suppose that a black box group B and a subgroup G are given, and G is a classical simple group of given characteristic. Suppose further that the algorithm in* Theorem 1.1 *produces an output $\mathcal{S}^*, \lambda$ for G. Then it is possible to decide by a deterministic algorithm whether any given element of B lies in G, $\mathrm{N}_B(G)$, or $G \cdot \mathrm{C}_B(G)$. The time requirement is $O(\mu q^{3/2}N + N^{5/2})$, $O(\mu q^{3/2}N^2 + N^{7/2})$, and $O(\mu q^{3/2}N^2 + N^{7/2})$, respectively.*

Theorem 1.1 can be extended to almost simple groups:

Corollary 1.4 *There is a Las Vegas algorithm which, when given a black box group $G = \langle \mathcal{S} \rangle$ known to be isomorphic to a group lying between G_0 and $\mathrm{Aut}\,G_0$, where G_0 is a classical simple group defined on a vector space over a field of given characteristic, finds a generating set $\mathcal{S}^*$ of G_0, a vector space V underlying G_0 (or vector spaces when $G_0 \cong \mathrm{PSL}(d,q)$ or $\mathrm{P\Omega}^+(8,q)$, and V is their direct sum), and a monomorphism $\lambda \colon G \to \mathrm{P\Gamma L}(V)$, such that $g\lambda$ and $h\lambda^{-1}$ can be found by straight-line programs of length $O(N)$ for any given $g \in G$ and $h \in G\lambda$. The running time is the same as in* Theorem 1.1.

An analogue of the algorithm of Corollary 1.4 also handles central extensions of almost simple groups $G = \langle \mathcal{S} \rangle$. Alternatively, we can consider $G/\mathrm{Z}(G)$ as a black box group, since membership testing in $\mathrm{Z}(G)$ (hence testing whether $g = 1$ in $G/\mathrm{Z}(G)$) is possible in $O(\mu|\mathcal{S}|)$ time: $z \in \mathrm{Z}(G)$ if and only if $[z, \mathcal{S}] = 1$. Consequently, the algorithm for Corollary 1.4 can be applied directly, with running time a factor $|\mathcal{S}|$ slower than the time for Theorem 1.1 (because of the slower black box group test for $g = 1$).

Since the orders of exceptional groups of Lie type are polynomial in the underlying field size q, these groups can be handled by brute force in time polynomial in q and the input length N. Hence algorithms analogous to Theorem 1.1 and Corollaries 1.2–1.4 exist for exceptional groups, with running time $O(\mu q^c N^c)$ for an absolute constant c (in fact, $c \leq 248$).

Another application is the combination of Theorem 1.1 with the methods of Beals and Babai [BeB]. They consider black box groups equipped with oracles to handle elementary abelian groups involved in G and discrete logarithms for the finite fields involved in G. Moreover, a superset of the primes occurring in $|G|$ is given. Then they find $|G|$, a composition series and Sylow subgroups of G by a Las Vegas algorithm, in time polynomial in $\nu(G)$ and the input length, where $\nu(G)$ is the smallest ν such that all nonabelian composition factors of G have faithful permutation representations of degree at most ν. As pointed out in [Be, Section 5], the term $\nu(G)$ in the running time can be replaced by an upper bound on the time requirement for handling the nonabelian composition factors of G. Since alternating groups can be handled in Las Vegas polynomial time in terms of the input length [BeB, BLNPS], and exceptional groups of Lie type have polynomial order in terms of the underlying field, we immediately deduce the following:

Corollary 1.5 *Suppose that a black box group $G = \langle \mathcal{S} \rangle$ is given, together with a list of primes that contains all prime divisors of $|G|$, and oracles for handling elementary abelian groups involved in G and discrete logarithms. Then there is a Las Vegas algorithm that computes $|G|$ and a composition series for G in time polynomial in the input length and q, where q is the size of the largest field involved in the definition of the Lie type composition factors of G.*

Sylow subgroups can also be found in the situation of the corollary. However, it is not possible only to quote the existing literature (as in [BeB]), since that depends heavily on permutation representations not available here. We note that a recent improved version [BaB] contains a Monte Carlo algorithm that finds black box representations for all nonabelian composition factors of $G = \langle \mathcal{S} \rangle$ in time polynomial in the input length, without using the linear algebra and discrete logarithm oracles. However, in order to find a composition series and determine the order of G, the oracles are still needed and the time requirement is of the same form as stated in the corollary.

Permutation group consequences require some background (cf. Section 8). We will be concerned with *nearly linear time* algorithms for permutation groups: randomized algorithms for computing with a given group $G = \langle \mathcal{S} \rangle \leq S_n$ in $O(|\mathcal{S}|n \log^c |G|)$ time for some constant c (in particular, in $O(|\mathcal{S}|n \log^{c'} n)$ time if G has a base of size $\log^{c''} n$; this base size is automatic for the classical groups considered here). The following corollary is covered by Theorem 1.1. Its first proof (with a Monte Carlo algorithm) was given by Morje [Mo1], where it was an essential ingredient for nearly linear time Sylow algorithms for permutation groups with no composition factor of exceptional Lie type.

Corollary 1.6 *There is a nearly linear time Las Vegas algorithm which, when given* $G = \langle S \rangle \leq S_n$ *known to be isomorphic to a classical simple group other than a 3–dimensional unitary group, finds the characteristic and Theorem 1.1(i-v).*

In contrast to Morje's result, Corollary 1.6 is Las Vegas because in the proof of Theorem 1.1 we construct a short presentation for the input group. The 3–dimensional unitary groups are excluded since it is not known whether they have a presentation of length polylogarithmic in their order (cf. [BGKLP]).

Although there is a large library of nearly linear algorithms for permutation groups (see [Ser1]), a major drawback is that almost all of these algorithms are Monte Carlo. Namely, a base and strong generating set are the basic data structures for computing with permutation groups and the starting points of most algorithms, but a Monte Carlo method is used for their nearly linear time computation [BCFS]. As a consequence of Corollary 1.6, we have the following result:

Theorem 1.7 *There is a nearly linear Las Vegas algorithm which, when given* $G = \langle S \rangle \leq S_n$ *having no composition factor isomorphic to an exceptional group of Lie type or a 3–dimensional unitary group, finds a base and strong generating set.*

We note that the hypothesis of Theorem 1.7, that G has no exceptional composition factors, can be checked by a nearly linear Monte Carlo algorithm.

As pointed out in [Ser2], all currently known randomized nearly linear Monte Carlo algorithms can be modified so that after an initial base and strong generating set computation, the rest of the algorithm is deterministic or Las Vegas. Thus, *for the groups described in* Theorem 1.7, *we can upgrade the entire nearly linear time library to Las Vegas.*

At a first glance, it may be surprising that black box recognition algorithms can be applied for the construction of strong generating sets. The method of Theorem 1.7 differs significantly from the traditional point-stabilizer chain constructions [Si1, Si2]; in the algorithm presented in **8.5**, by the time we have found $|G|$ we have also constructed a composition series for G. In this respect, the algorithm resembles the parallel (NC) handling of permutation groups [BLS1] and the current fastest deterministic algorithms for computing strong generating sets [BLS2, BLS3]. Part of the black box recognition process for simple groups is a construction of a short presentation, which plays an important role in the verification of the correctness of the strong generating set construction as well.

In fact, a constructive short presentation for black box classical simple groups also has the following consequence for permutation groups, in view of [BGKLP]:

Corollary 1.8 *There is a nearly linear Las Vegas algorithm which, when given $G \leq S_n$ with no exceptional Lie type or 3-dimensional unitary composition factors, computes a presentation of length $O(\log^3 |G|)$ for G.*

Theorem 1.1 can also be used to speed up the nearly linear Sylow subgroup computations in simple groups in [Mo1]. Given an arbitrary permutation representation of a simple classical group G, Morje first constructs the permutation action of the group on the (singular) points of the natural module, and then a matrix representation for G. Theorem 1.1 constructs the matrix representation directly, eliminating the need for passing to the second permutation representation.

Built into Theorem 1.1 is yet another corollary: *a Monte Carlo algorithm that decides whether or not a given simple black box group is isomorphic to some classical group defined over a field of given characteristic.* A Monte Carlo rather than a Las Vegas algorithm seems called for here: it is difficult to imagine that there could be an efficient algorithm that decides, with certainty, that G is *not* isomorphic to $\mathrm{PSL}(d, q)$ for any d and q, without a complete structural exploration of G as in Corollary 1.5.

The stated results suggest the wide applicability of Theorem 1.1. However, as in [CFL] the required background is much less than in other papers of a similar nature: the classification of the finite simple groups will not be used in this paper—except in Theorem 1.7 and its corollary. We have replaced group-theoretic methods with geometric ones. Of course, when our results are used elsewhere the classification will, presumably, be needed in order to obtain our hypotheses. Nevertheless, the present approach, and those of [CFL, CLG2], have the advantage of dispensing with difficult group theory and aiming more directly at the interesting algorithmic aspects of questions such as these.

We require a great deal of detailed information concerning the classical groups. While we have outlined some of this information when needed, we have had to assume that readers will be familiar with significant amounts of background contained, for example, in [As2, KL, Ta]. This need should not be surprising, since this information provides the main tool available here.

One main difference between our treatment and that of [CFL] is our greater use of a geometric rather than a linear algebra point of view (cf. [Ka3, Ka4, Mo1, Mo3]); another is that [CFL] includes a significant amount of implementation details; yet another is that the prime field seems to allow rather different types of ideas in [CFL] (which have now been upgraded to arbitary q in [BCFL]). The algorithm in [CFL] leads to better timing estimates than the one presented here in the case $PSL(d, 2)$ studied in that paper. As in [CFL], we cannot use Gaussian elimination as was done in [CLG2]: that reference already had the "correct" vector space within which to compute. On the other hand, Gaussian elimination enabled [CLG2] to contain a more efficient and practical algorithm, which has already been implemented in GAP and MAGMA. In view of these two papers, we view handling the orthogonal, symplectic and unitary groups as the most significant aspect of our treatment. See **9.1** for further comments concerning the approaches in those papers and this one.

We are indebted to Prabhav Morje and his thesis research [Mo1]. The present research began as an attempt by the first author to simplify some of the complicated orthogonal group considerations in [Mo1]. This led first to Lemma 4.8, then to **4.2.1**, and finally to the rest of this paper.

Overview of the Proofs

For the proof of Theorem 1.1 we also need a slight variation for our recursive purposes, in which groups of linear transformations arise rather than projective groups:

Theorem 1.1′ *There is a Las Vegas algorithm which, when given a black box group $G = \langle S \rangle$ such that $G = G'$, $G/Z(G)$ is simple, and G is isomorphic to a classical linear group defined on a d–dimensional vector space over a field $GF(q)$, produces the same conclusions as Theorem 1.1, but this time with a monomorphism $G \to SL(d, q)$ mapping onto the appropriate classical group. Moreover, $Z(G)$ and $Z(G)\lambda$ can be found using straight-line programs as in Theorem 1.1(iii,iv). The running times of the algorithms are the same as in Theorem 1.1.*

Note that the cyclic group $Z(G)$ is not part of the input, although we know its order (which divides $(d, q - 1)$). A very slightly more general result than Theorems 1.1 and 1.1′ can be obtained, in which G is assumed to be quasisimple with $G/Z(G)$ a simple classical group. Once again a constructive isomorphism is obtained for $G/Z(G)$. The proof is almost the same as for the above theorems (see the remarks following Corollary 1.4 above).

Section 2 briefly introduces black box groups and some of their properties, but we do *not* use most of the known algorithms for black box groups. The proof of Theorems 1.1 and 1.1' occupies Sections 3–7. As already mentioned, the details of our treatment of the different classical groups vary too much for a uniform treatment. Instead, we have arranged each section into subsections, tailored so that the similarities can be emphasized even when details differ. An outline of each of these subsections is as follows.

1. Background concerning the groups: transvections and root elements, $Q(\alpha) = O_p(G_\alpha)$ for a 1–space (or hyperplane) α, probabilistic generation results.

2. Find a transvection or a long root element. Implicitly study the resulting conjugacy class of G, but without ever computing more than a small portion of this class.

3. Find G'_α and $Q(\alpha) = O_p(G'_\alpha)$, even though α itself is not available.

   Conjugates $Q(\alpha)^g$ are viewed as "points" (1–spaces) of the target vector space V. Effective and deterministic transitivity is obtained for the action of $Q(\alpha)$ on a set of points, even though we do not have the vector space and cannot list more than a small number of these points within our time constraints.

   This transitivity is used to find a complement L to $Q(\alpha)$ in G'_α. Now recursion is invoked for L. The remainder of the algorithm in **4** and **5** is deterministic.

4. Introduce linear algebra internally for a specific quotient $\bar{Q}(\alpha)$ that is a vector space over $\mathrm{GF}(p)$.

5. An algorithm decides, for each point, how it should be labeled as a 1–space of V. This produces a bijection from $Q(\alpha)^G$ to a set of 1–spaces of V, which is then converted into a map taking any given element of G to a matrix.

   We also provide straight-line programs for Theorems 1.1(iii,iv) and 1.1'(iii,iv); recursive use of the latter was critical for setting up the data structure needed for constructing V.

   This is the main part of the algorithm: most of the preceding parts can be viewed as preprocessing steps.

6. Determine the total timing, and deal with small–dimensional cases.

7. In **1–6** we assumed that we knew the dimension, the underlying field, and the type of G. With this extra information, the algorithm is Las Vegas. The entire algorithm starts by finding at most 7 possibilities for these parameters, and then for each the procedure in **1–6** is called. If an output is returned, we try to verify that our assumption on the isomorphism type of G was correct by checking that G satisfies a presentation of the target group.

Names of subsections **1–6** in Sections 3–6 parallel the above list, while **7** is in Section 7. For a simplified though somewhat different algorithm for the special case $PSL(2, q)$ of Theorem 1.1, see **3.6.1**. However, both in this special case and in general readers may prefer to simplify the theorems and proofs by handling small-dimensional examples by brute force methods, and by reading Theorem 1.1 as saying that the algorithms run in time polynomial in N and q.

The family of groups $PSL(V)$, in some ways easier than the other families, forces us to deal with what amount to two different permutation representations simultaneously: the actions on 1–spaces and on hyperplanes of the target vector space (Section 3). In [CFL] playing off these two representations was an important tool, here it is seen as more of a nuisance that does not arise for the other classical groups. The orthogonal groups (Section 4) differ from the other families in not having transvections, which makes it somewhat harder to work with "points". Symplectic groups (Section 5) are the easiest case when the characteristic is odd, but involve difficulties in characteristic 2 caused by their behaving like orthogonal groups. Unitary groups (Section 6) require the fewest special cases. The symplectic and unitary cases are especially different from the case $PSL(d, q)$ in the behavior of their transvection groups: the commutator relations are quite different and lead to methods that are new in many details.

Section 7 ties all of this together, and contains proofs of Corollaries 1.2–1.4. Finally, the proofs of Theorem 1.7, and of Corollaries 1.6 and 1.8, are contained in Section 8. Among other things, in that section we introduce background for the nearly linear time constraint.

Theorems 1.1 and 1.7 were announced at the Oberwolfach Computational Group Theory meeting in June, 1997, as well as in [Ka5].

2 Preliminaries

2.1 Notation

Let X be a group. Its largest normal p–subgroup is $O_p(X)$ for any given prime p. If $X = X'$ and $X/\mathrm{Z}(X)$ is a nonabelian simple group, then X is called *quasisimple*. If $X \leq G$ and $g \in G$ then g^X is $\{g^x \mid x \in X\}$. If X and Y are subgroups of a group G such that $G = XY$, $X \trianglelefteq G$ and $X \cap Y = 1$, we write $G = X \rtimes Y$ (a *semidirect product*) and call Y a *complement* to X in G.

If $a, b \in G$ then $[a, b] = a^{-1}b^{-1}ab$. For $A, B \subseteq G$, write $[A, B] = \langle [a, b] \mid a \in A,\ b \in B\rangle$.

If X acts on a vector space V, we will view X and V as subgroups of the resulting semidirect product $V \rtimes X$. Then $\mathrm{C}_V(X)$ is the space of fixed vectors, and $[V, X] = \langle vx - v \mid v \in V,\ x \in X\rangle$.

2.2 Black box groups

A *black box group* G is a group whose elements are encoded as 0-1 strings of uniform length N, and the group operations are performed by an oracle (the "black box"). Given strings representing $g, h \in G$, the black box can compute strings representing gh and g^{-1}, and decide whether $g = h$. Note that $|G| \leq 2^N$: we have an upper bound on $|G|$. We always let μ be as in Theorem 1.1: an upper bound on the time for each multiplication or inversion or equality test within our group G. Clearly $\mu \geq N$.

Each string represents at most one element of G, and the same element of G may be encoded by different strings.

2.2.1 Listing subgroups

We will often need to list all elements of small subgroups H of a black box group G.

Lemma 2.1 *Let $H = \langle T\rangle \leq G$ be an elementary abelian p-group. Then H can be listed in $O(\mu|H||T|)$ time. Moreover, in the same time, a subset of T which is a $\mathrm{GF}(p)$–basis for H can be chosen, and the coefficients of each element as a linear combination of the basis can be computed.*

Proof. Let $0 \leq j \leq |T|$, and let L_j be a listing of the group generated by the first j generators of H. If the $(j + 1)$st generator g of T does not occur in

11

L_j then we add g to the basis and compute $L_{j+1} := L_j \cup L_j g \cup \cdots \cup L_j g^{p-1}$. Otherwise, we define $L_{j+1} := L_j$ and discard g. Note that this procedure also computes how the elements of H are written as linear combinations of the basis vectors. $\square$

There are other general situations in which one can list small nonabelian groups. We will only need to consider a few very special instances: **6.3.2**, **6.6.1** and **6.6.2**.

2.2.2   Random elements

We will need random elements of black box groups. We say that an algorithm outputs an ε–*uniformly distributed element* x in a group G if $(1 - \varepsilon)/|G| <$ $\mathrm{Prob}(x = g) < (1 + \varepsilon)/|G|$ for all $g \in G$. *Nearly uniform* means ε–uniform for some $\varepsilon \leq 1/2$; we also use the terms "ε–uniformly distributed" and "ε–uniformly distributed random" interchangeably. The following fundamental result is due to Babai.

Theorem 2.2 [Ba1] *Let c and C be given positive constants. Then there is a Monte Carlo algorithm which, when given a black box group G of order at most M and any set of generators $\mathcal{S}$ of G, sets up a data structure for the construction of ε–uniformly distributed elements for $\varepsilon = M^{-c}$, at a cost of $O(\log^5 M + |\mathcal{S}| \log \log M)$ group operations. The probability that the algorithm fails is $\leq M^{-C}$.*

If the algorithm succeeds, it permits the construction of ε–uniformly distributed, independent random elements of G at a cost of $O(\log M)$ group operations per element.

In view of Theorem 2.2, it is possible to express the parameter ξ as a polynomial function of N and μ in Theorem 1.1. We chose to introduce the notation ξ to ensure that the running time estimates can be adapted easily to other constructions of random elements in G. *We assume that $\xi \geq \mu|\mathcal{S}|$, since* it is presumed that each generator should be involved in the construction of a random element. Of course, in all subgroups of the input group G we have $M \leq 2^N$; however, when working with small subgroups of G, we have smaller values of M, and hence faster algorithms. Note also that using Theorem 2.2, the construction of subsequent random group elements is much cheaper than the construction of the first one. A simpler and more practical heuristic algorithm for finding random group elements is given in [CLMNO].

In this paper, we assume the availability of ε-uniformly distributed, independent random elements of G for some $\varepsilon \leq 1/2$. In fact, in the probability analysis, we assume the availability of uniformly distributed, independent random elements of G. However, since our algorithm ends by checking a presentation for G, it will return a correct output no matter what method of random generation is used. Using a less efficient method may only increase the probability of failure for the algorithm.

2.2.3   Orders of elements

Another task we frequently encounter is to establish properties of orders of group elements. Although the details will vary, we can usually reduce the task to the following basic problem: given $g \in G$ and an integer k, decide whether $|g|$ divides k. This can be done by computing g^k (by repeated squaring and using the binary expansion of k, requiring $O(\log k)$ black box group operations), and compare g^k and 1. We emphasize that in general we will not attempt to compute the order of elements of G, or factor the integers k occurring in this situation. This is in contrast of the algorithms in [CFL, CLG2], and avoids the difficulties of factoring large integers. In **2.4** we will comment further on the integers k occurring as above.

The only cases when we search for group elements with order exactly k occur with $k \in \{q-1, (q-1)/(2, q-1)\}$. As preprocessing, the prime factorization of these numbers can be computed in $O(q)$ time. Then testing $|g| = k$ takes $O(\mu \log^2 q)$ time: for each of the at most $\log q$ prime divisors r of k, and also for $r = 1$, compute the k/rth power of g via repeated squaring.

2.2.4   The parameters μ and ξ

Of course, μ depends on the actual representation of the black box group. An important example is when G is represented as a group of $a \times a$ matrices over a finite field F. We will assume that field operations can be carried out in constant time (cf. **2.3**), so that group operations can be performed in $O(a^3)$ time (of course, they could be sped up by using faster multiplication algorithms). Another example, important for Theorem 1.7 and Corollaries 1.6 and 1.8, is when G is represented as a small–base permutation group. We will see in Section 8 that μ is $O(\log^c n)$ for a small–base permutation group of degree n.

The parameter ξ was defined in Theorem 1.1. One estimate appears in Theorem 2.2. It is clear that $\xi = \xi(G)$ depends on G. We will always assume that this dependency is monotonic: *if $L < G$, then $\xi(L) \le \xi(G)$.*

2.2.5   Straight-line programs

Given $G = \langle \mathcal{S} \rangle$, a *straight-line program of length m* from $\mathcal{S}$ to some $g \in G$ is a sequence of expressions $(w_1, \ldots, w_m)$ such that for each i one of the following holds:

- w_i is a symbol for some element of $\mathcal{S}$,
- $w_i = (w_j, -1)$ for some $j < i$, or
- $w_i = (w_j, w_k)$ for some $j, k < i$,

such that, if the expressions are evaluated, then the value of w_m is g. Here, $(w_j, -1)$ is evaluated as the inverse of the evaluated value of w_j, and (w_j, w_k) is evaluated as the product of the evaluated values of w_j and w_k. Thus, a straight-line program is really a sequence of elements of the free group on $\mathcal{S}$. While it is common to think of the terms w_i as elements of G, we save significant time by not writing each w_i as, for example, a $d \times d$ matrix when we do not actually need the specific intermediate matrices later. This will allow us to mirror a straight-line program using a homomorphism, and only *then* evaluate the members of the sequence as group elements.

A straight-line program actually *proves* that $g \in G$, and straight-line programs of length $\Theta(m)$ can reach some elements g at distance $\Theta(2^m)$ from the identity in the Cayley graph of G with respect to $\mathcal{S}$. In fact, Babai and Szemerédi [BSz] showed that, for any $G = \langle \mathcal{S} \rangle$ and $g \in G$, there exists a straight-line program from $\mathcal{S}$ to g using at most $(\log |G| + 1)^2$ multiplications and inversions. In Theorem 1.1, we provide a constructive version of this for classical matrix groups, though with a weaker bound on the length of such a program.

2.3   Fields

The algorithm for Theorem 1.1 starts by finding the value of q (cf. **7.2.1**). After that, as a preprocessing step, we have to construct a field $\mathbb{F} = \mathrm{GF}(q)$, $q = p^e$. The elements of this field will be used as entries of the output matrices in Theorem 1.1(ii). Moreover, during the course of the algorithm, we shall construct sets of endomorphisms of elementary abelian groups which are isomorphic to $\mathrm{GF}(q)$. We shall always identify these fields with $\mathbb{F}$.

For the construction of $\mathbb{F}$, we need a linear transformation of order $q - 1$ on the vector space $\mathrm{GF}(p)^e$. We shall obtain such a linear transformation as a side result when processing the group $\mathrm{PSL}(2, q)$ or $\mathrm{SL}(2, q)$. This is sufficient, regardless of the isomorphism type of the input group G, since the recursion of the algorithm for Theorem 1.1 always arrives at a 2–dimensional special linear group before the need to use $\mathbb{F}$ arises.

An alternative way to find a linear transformation of order $q-1$ is to write all $e \times e$ companion matrices ρ over $\mathrm{GF}(p)$, find one of order $q - 1$, and let $\mathbb{F} := \{0\} \cup \langle \rho \rangle$. There are p^e companion matrices, and the order of one matrix can be tested in $O(e^3 \log^2 q)$ time: for each of the at most $\log q$ prime divisors r of $q - 1$ and also for $r = 1$, compute the $(q-1)/r$th power of ρ via repeated squaring in $O(e^3 \log q)$ time. (A prime factorization of $q - 1$ can be computed in $O(q)$ time.) Here and in the sequel, we shall assume that field operations in the prime field $\mathrm{GF}(p)$ take constant time; in particular, we store a list of inverses for $\mathrm{GF}(p)^*$.

Throughout this paper, the symbol ρ will always refer to such an element of $\mathbb{F}$.

However we have obtained ρ, as a second preprocessing step in $O(qe^2 \log q)$ time we construct a Zech logarithm table for $\mathbb{F}$ [Con]. This is a list of length $q-1$ which, for the above generator ρ, stores the values $\rho^i + 1$, $0 \le i < q-1$, as powers of ρ or 0. We list the powers of ρ in $O(e^3 q)$ time and order them lexicographically, considered as e^2-long sequences of integers from $\{0, 1, \ldots, p-1\}$, in $O(e^2 q \log q)$ time. Then, for each ρ^i, use binary search to find $\rho^i + 1$ among the elements of this list in $O(e^2 \log q)$ time. The entries of the matrices in the image of the homomorphism λ constructed in Theorem 1.1(ii) are symbols (0 and powers of ρ) from the Zech logarithm table.

After the Zech logarithm table has been constructed, we consider the elements of $\mathbb{F}$ only as symbols ρ^i, $0 \le i < q - 1$, and 0. Then multiplication is just the addition of exponents of ρ: $\rho^i \cdot \rho^j = \rho^{i+j}$. Addition is a lookup in our table, followed by a multiplication: $\rho^i + \rho^j = \rho^j(\rho^{i-j} + 1)$. Thus, *throughout this paper we assume that field operations in $\mathbb{F}$ can be carried out in constant time.*

The set $\{\rho^i \mid 0 \le i < e\}$ is a $\mathrm{GF}(p)$–basis of $\mathbb{F}$. As a third preprocessing step, we compute and store the coefficients of each $\zeta \in \mathbb{F}$ in this basis. This can be done by computing an $e \times e$ matrix M with entries $M(i, j) := \rho^{(i-1)p^{j-1}}$, and then also M^{-1}. Then the coefficients of ζ are the coordinates of the vector $(\zeta, \zeta^p, \ldots, \zeta^{p^{e-1}})M^{-1}$ [LN, Ch. 2]. This computation takes $O(qe^2)$ time.

In each of Sections 3–6 we will also need to consider certain elementary abelian p-subgroups or sections Q of G as vector spaces over $\mathbb{F}$. Multiplication by the elements of the Zech logarithm table is not possible, so we shall construct other copies of $GF(q)$ (one for each Q we consider) which act as endomorphism rings of the sections Q involved. These fields have to be identified with $\mathbb{F}$; in some cases it will be straightforward, in others we shall use the following lemma.

Lemma 2.3 *Let A be an elementary abelian black box p-group, with parameter μ as before. Let E be a subset of the endomorphism ring of A such that $E \cong GF(q)^*$, where $q = p^e$ and E acts homogeneously on A. Suppose further that E is given as a list of $q - 1$ procedures such that for all $1 \neq a \in A$ and all $\varepsilon \in E$, $a\varepsilon$ can be computed in $O(\nu)$ time. Then, in $O(\mu q \log q + \nu q e)$ time, we can find $\zeta \in E$ such that the mapping $\zeta^k \mapsto \rho^k$, $0 < i < q$, defines a field isomorphism from E to $\mathbb{F} - \{0\}$.*

Proof. Recall that we have stored the expression $\rho^e = \sum_0^{e-1} b_i \rho^i$ using some $b_i \in GF(p)$. Fix some $1 \neq a \in A$ and check for each $\varepsilon \in E$ whether $a\varepsilon^e = \sum_0^{e-1} ab_i\varepsilon^i$. A solution ε of this equation is appropriate as ζ. $\square$

In our applications, the "procedures" for computing $a\varepsilon$ will be conjugations by elements of some overgroup $B \rhd A$ or taking the product of a with a conjugate of a as in the proof of the following lemma.

As an additional technical difficulty, sometimes we have available only a (large) part of the field acting on an elementary abelian subgroup A. This will occur in the following situation. We have a listing of A, $|A| = q$, and also we have the coefficients of each $a \in A$ in terms of a fixed basis. We also have a set E' of $(q - 1)/2$ endomorphisms, given by conjugation with powers of some element $b \in B$, where $B \rhd A$, such that E' is cyclic of order $(q - 1)/2$.

Lemma 2.4 *Suppose that A and E' are given as described above. Then $q - 1$ procedures for E as in Lemma 2.3 can be found in $O(\mu q \log q)$ time.*

Proof. Let ε denote the endomorphism defined by conjugation with b, and let $a_1, \ldots, a_e$ be the given basis of A. First, we construct the $e \times e$ matrix M over $GF(p)$ corresponding to ε; the rows of M are the coordinate vectors of $a_i\varepsilon$. Then we fix $1 \neq a \in A$, with coordinate vector v_0, and construct the list L of the coordinate vectors v_k for $a\varepsilon^k$, $0 \leq k < (q - 1)/2$. Finally, we search for v_k such that $v_k + v_0 \notin L$. For such a k we have $E = E' \cup E'(\varepsilon^k + 1)$.

The matrix M can be constructed in $O(\mu q e)$ time, by identifying a_i^b in the listing of A for $1 \leq i \leq e$. The elements of L are the vectors $v_0 M^k$, computed in $O(q e^2)$ time. Then L can be ordered lexicographically in $O(e q \log q)$ time, and after that deciding by a binary search whether $v_j + v_0 \in L$ for a fixed j takes $O(e \log q)$ time. $\square$

We will always use only naive linear algebra: Gaussian elimination to solve systems of linear equations, and multiplication and inversion of $d \times d$ matrices in $O(d^3)$ time.

2.4 Primitive divisors and irreducible linear transformations

By a fundamental theorem of Zsigmondy [Zs], if p is prime and $m \geq 2$ then there is a prime dividing $p^m - 1$ but not $p^i - 1$ for $1 \leq i < m$, except when either $p = 2, m = 6$, or $m = 2$ and p is a Mersenne prime. Such a prime is called a *primitive prime divisor* of $p^m - 1$.

We will call an integer j a $\mathrm{ppd}^\#(p; m)$ if $j | p^m - 1$ and either $p = 2, m = 6$ and $21 | j$; $m = 2$, p is a Mersenne prime, and $4 | j$; or j is divisible by a primitive prime divisor of $p^m - 1$.

Any element $g \in G = \mathrm{GL}(d, p^e)$ of $\mathrm{ppd}^\#(p; ed)$–order is irreducible on $\mathrm{GF}(p^e)^d$. Moreover, $\mathrm{C}_{\mathrm{GL}(d,p^e)}(g) \cong \mathrm{GF}(p^{ed})^*$ and $\mathrm{N}_{\mathrm{GL}(d,p^e)}(\langle g \rangle)/\mathrm{C}_{\mathrm{GL}(d,p^e)}(g)$ is isomorphic to the Galois group of $\mathrm{GF}(p^{ed})$ over $\mathrm{GF}(p^e)$. From this it is easy to check that there are many such elements:

Lemma 2.5 [NP] *Exclude the cases $p = 2, ed = 6$ and $p = 3, ed = 2$. If $g \in G = \mathrm{GL}(ed, p)$ has $\mathrm{ppd}^\#(p; ed)$–order, then at least half of the elements of $\mathrm{C}_{\mathrm{GL}(d,p^e)}(g)$ have $\mathrm{ppd}^\#(p; ed)$–order. $\mathrm{GL}(d, p^e)$ has more than $|\mathrm{GL}(d, p^e)|/2d$ elements of $\mathrm{ppd}^\#(p; ed)$–order.*

We will need to find elements $g \in \mathrm{GL}(d, q)$ of $\mathrm{ppd}^\#(p; m)$–order for a variety of choices of m. We will abuse language and abbreviate this by using expressions such as $|g| = \mathrm{ppd}^\#(p; m)$, or $|g| = p \cdot \mathrm{ppd}^\#(p; m)$ if $|g|$ is to be p times a $\mathrm{ppd}^\#(p; m)$.

Lemma 2.6 *If the case $p = 2$, $m = 6$ is excluded, then a group element g such that $g^{p^m - 1} = 1$ and $g^{\prod_1^{m-1}(p^k - 1)} \neq 1$ has $\mathrm{ppd}^\#(p; m)$–order unless $m = 2$, $p \equiv 3 \pmod 4$, and $|g|$ is a power of 2. Hence, if $m > 2$ and $g^{p^m - 1} = 1$, $O(m^2 \log p)$ multiplications can be used to test whether g has $\mathrm{ppd}^\#(p; m)$–order.*

Proof. Let $x := \prod_1^{m-1}(p^k - 1)$. For any odd prime r dividing $p^m - 1$, if r is not $\mathrm{ppd}^\#(p; m)$ then $(p^m - 1)_r = r \cdot (p^{m/r} - 1)_r \leq (p^{m/r} - 1)_r \cdot (p^{2m/r} - 1)_r \leq x_r$.

Similarly, $(p^m - 1)_2 > x_2$ if and only if $m = 2$ and $p \equiv 3 \pmod 4$. Hence any odd prime divisor of $|g^x|$ is a $\mathrm{ppd}^{\#}(p; m)$–number and if $|g^x|$ is a power of 2 then $m = 2$ and $p \equiv 3 \pmod 4$. $\square$

We note that [NP] contains a procedure with running time $O(m^3 \log m \log^3 p + \mu m \log p)$ that checks the $\mathrm{ppd}^{\#}(p; m)$ property of a group element of order dividing $p^m - 1$.

We will use the following when dealing with all of the classical groups:

Lemma 2.7 *Suppose that $g \in \mathrm{GL}(d, p^e)$ has $\mathrm{ppd}^{\#}(p; ed)$–order. Then $\{v^{g^i} \mid 0 \le i < ed\}$ is a $\mathrm{GF}(p)$–basis of $\mathrm{GF}(p^e)^d$ and $\{v^{g^i} \mid 0 \le i < d\}$ is a $\mathrm{GF}(p^e)$–basis of $\mathrm{GF}(p^e)^d$ for any nonzero $v \in \mathrm{GF}(p^{ed})$.*

Proof. The minimal polynomial of g over $\mathrm{GF}(p)$ has degree ed. The span of g is a field $\mathrm{GF}(p^{ed})$, and $\mathrm{GF}(p^{ed})^*$ acts regularly on $\mathrm{GF}(p^e)^d - \{0\}$, so that $\{v^{g^i} \mid 0 \le i < d\}$ is linearly independent over $\mathrm{GF}(p^e)$ and hence is a basis of $\mathrm{GF}(p^e)^d$. $\square$

Also $\mathrm{ppd}^{\#}(p; 1)$. In **3.1.4**, **4.1.4**, **4.2**, **4.6.2** and **5.2.2** we will need to consider factors of $p - 1$. Therefore, it is convenient to extend the notion of primitive divisors by calling an integer j a $\mathrm{ppd}^{\#}(p; 1)$ if $j|p - 1$, $j > 2$, and j is not a power of 2 if $p - 1$ is not.

2.5   Probability estimates

We shall mostly use elementary probability estimates. The only exception is a slight extension of Chernoff's bound [Che]:

Lemma 2.8 [BCFLS] *Let $Y_1, \ldots, Y_t$ be not necessarily independent, $0, 1$ valued random variables with the property that, for some r and each i, the conditional probability $\mathrm{Prob}(Y_i = 1 \mid Y_1 = x_1, \ldots, Y_{i-1} = x_{i-1}) \ge r$ for all 0-1 sequences $(x_1, \ldots, x_{i-1})$. Then, whenever $0 < \varepsilon < 1$,*

$$\mathrm{Prob}\Big(\sum_{i=1}^{t} Y_i \le (1 - \varepsilon)rt\Big) \le e^{-\varepsilon^2 rt/2}.$$

We shall apply Lemma 2.8 in two situations.

(1) Suppose that at least $r|H|$ elements of a group H satisfy a certain property $\mathcal{P}$. If we choose a sequence $(y_1, \ldots, y_m)$ of uniformly distributed, independent random elements of H and define that the random variable $Y_i =$

1 if and only if y_i satisfies $\mathcal{P}$, then $\sum Y_i$ simply measures the number of y_i satisfying $\mathcal{P}$.

(2) The second application is more complicated. Suppose that H is an elementary abelian p-group with $|H| = p^m$, and $(y_1, \ldots, y_s)$ is a sequence of elements from H with the following property: for all $i \leq s$, if $\langle y_1, \ldots, y_{i-1} \rangle \neq H$ then $\mathrm{Prob}(y_i \notin \langle y_1, \ldots, y_{i-1} \rangle) \geq r$. Then, if we define $Y_i = 0$ if and only if $\langle y_1, \ldots, y_{i-1} \rangle \neq H$ and $y_i \in \langle y_1, \ldots, y_{i-1} \rangle$, the condition $\sum Y_i \geq m$ means that the subgroup chain generated by the initial segments of the sequence $(y_1, \ldots, y_s)$ increases m times, so that *the y_i's generate H*.

3 Special linear groups: $\mathrm{PSL}(d, q)$

In this section we will prove Theorems 1.1 and 1.1′ when G is $\mathrm{PSL}(V) = \mathrm{PSL}(d, q)$ or $\mathrm{SL}(V)$. As before we write $q = p^e$ with p prime.

3.1 Properties of G

We presuppose familiarity with the classical groups. See [As2, Ta, KL] for introductions to these groups. Here we will state those properties needed later, in general not giving proofs or references for straightforward calculations.

3.1.1 Transvections

Let $\alpha = \langle u \rangle$ be a *point* (1–space) of V and let $0 \neq f \in V^*$ (the dual space of V), where $f(u) = 0$. A *transvection* with *axis* $W = \ker f$ and *center* (or *direction*) α is either 1 or a linear transformation of the form $t \colon v \mapsto v + f(v)u$. We will usually abuse notation and write expressions such as "$\alpha \in W$" instead of "$\alpha \subseteq W$" when discussing points: many ideas in this paper are geometric.

The corresponding *transvection group* is

$$T = T(\alpha, W) = \{v \mapsto v + kf(v)u \mid k \in \mathrm{GF}(q)\} \cong \mathrm{GF}(q)^+. \qquad (3.1)$$

Clearly, if $1 \neq t \in T$ then $\alpha = [V, t] = [V, T]$ and $W = \mathrm{C}_V(t) = \mathrm{C}_V(T)$.

There are $(q^d - 1)(q^{d-1} - 1)/(q - 1)$ nontrivial transvections. In particular, this number is $q^2 - 1$ when $d = 2$, in which case there are $(2, q - 1)$ conjugacy classes of such transvections in $\mathrm{SL}(2, q)$, each of size $(q^2 - 1)/(2, q - 1)$.

3.1.2 Commutator relations

Consider transvection groups $T_i = T(\alpha_i, W_i)$, $i = 1, 2$. There are five possibilities for the way T_1 and T_2 interact:

(i) $\alpha_1 = \alpha_2$ or $W_1 = W_2$, $[T_1, T_2] = 1$.

(ii) $\alpha_1 \in W_2$, $\alpha_2 \in W_1$, $[T_1, T_2] = 1$.

(iii) $\alpha_1 \in W_2$, $\alpha_2 \notin W_1$, and $[t_1, T_2] = [T_1, t_2] = T(\alpha_1, W_2)$ whenever $1 \neq t_1 \in T_1$, $1 \neq t_2 \in T_2$. Moreover, $[T_1, T_2]$ *commutes with both* T_1 *and* T_2.

(iv) $\alpha_2 \in W_1$, $\alpha_1 \notin W_2$. Then $[T_1, T_2]$ behaves as in (iii).

20

(v) $\alpha_1 \notin W_2$, $\alpha_2 \notin W_1$, and $\langle t_1, T_2 \rangle = \langle T_1, T_2 \rangle \cong \mathrm{SL}(2, q)$ whenever $1 \neq t_1 \in T_1$. Here, $\langle T_1, T_2 \rangle$ preserves the decomposition $V = \langle \alpha_1, \alpha_2 \rangle \oplus (W_1 \cap W_2)$, inducing $\mathrm{SL}(2, q)$ on the first summand and 1 on the second one.

These, and similar assertions in later sections, are easy to check using either linear algebra or the geometry of V. They are special cases of the commutator relations for groups of Lie type [Ca, pp. 76-77].

3.1.3 Q and $Q(\alpha)$

If W is a hyperplane of V then its set–stabilizer G_W can be written

$$G_W = Q \rtimes G_{W\gamma} \quad \text{for any point } \gamma \notin W. \tag{3.2}$$

Here, $Q \cong \mathrm{GF}(q)^{d-1}$ consists of all transvections

$$r(u) = \begin{pmatrix} I & O \\ u & 1 \end{pmatrix} \quad \text{for } u \in \mathrm{GF}(q)^{d-1}. \tag{3.3}$$

with axis W (using a basis of V that starts with a basis of W), and $G_{W\gamma}$ acts irreducibly on Q as a subgroup of $\mathrm{GL}(d-1, q)$ containing $\mathrm{SL}(d-1, q)$:

$$r(u)^l = r(u^l) \quad \text{for } l \in \mathrm{SL}(W). \tag{3.4}$$

Moreover, $\mathrm{C}_G(Q) = Q \times \mathrm{Z}(G)$. Here, Q is regular on the set of points of V not in W: *there is a unique element of Q taking any given point of this sort to any other such point.* This will be crucial for **3.5.1**. By **3.1.2**(iii,iv), *if t is any nontrivial transvection with center on W and axis not W, then $[Q, t]$ is the transvection group with axis W having the same center as t.*

Dually, for any point α of V, let $Q(\alpha)$ denote the group of all p–elements inducing 1 on V/α; this is the group of all transvections with center α. With respect to a suitable basis, matrices for $Q(\alpha)$ look like the transposes of the ones just given for Q.

We need one further property of Q:

Lemma 3.5 *Let $d \geq 3$. Each element $h \in G_W$ induces on Q a $\mathrm{GF}(q)$–linear transformation $\tilde{h}$ whose determinant is a dth power. If h has matrix $\begin{pmatrix} M & O \\ * & \rho \end{pmatrix}$ for a generator $\rho = (\det M)^{-1}$ of $\mathrm{GF}(q)^*$, then $\det \tilde{h}$ generates the group of dth powers within $\mathrm{GF}(q)^*$.*

3.1.4 Probabilistic generation

Throughout this paper we will need results concerning groups generated by transvections (or their analogues in orthogonal groups). These results are of two types: lists of irreducible subgroups generated by transvection groups, and the probable structure of a group generated by a very small number of randomly chosen transvections. In each section we will need slightly different results of these two sorts.

Theorem 3.6 [Mc1, Mc2] *For $d \geq 3$, each irreducible subgroup of $\mathrm{SL}(d, q)$ generated by transvection groups is one of the following:*

  (i) $\mathrm{SL}(d, q)$,

  (ii) $\mathrm{Sp}(2n, q)$ *in* $\mathrm{SL}(2n, q)$,

  (iii) $\mathrm{O}^{\pm}(2n, 2)$ *in* $\mathrm{SL}(2n, 2)$, *or*

  (iv) S_{2n+2} *or* S_{2n+1} *in* $\mathrm{SL}(2n, 2)$.

Lemma 3.7 *If $d \geq 4$ then, with probability $> (1-1/q)^5 \geq 1/2^5$, three nontrivial transvections generate a group that preserves a decomposition $V = U \oplus Z$ with $\dim U = 3$, inducing $\mathrm{SL}(3, q)$ on U and 1 on Z.*

Proof. For $i = 1, 2, 3$, let t_i be randomly and independently chosen transvections, with center α_i and axis W_i. Then $\alpha_1 \notin W_2$ and $\alpha_2 \notin W_1$ with probability $> (1-1/q)^2$. Next, $\alpha_3 \notin W_1 \cap W_2$ and $\alpha_3 \notin \langle \alpha_1, \alpha_2 \rangle$ with probability $\{[(q^d - 1)/(q-1)] - (q+1) - [(q^{d-2} - 1)/(q-1)]\}/\{(q^d - 1)/(q-1)\} > 1 - 1/q$. Finally, W_3 does not contain the point $\langle \alpha_1, \alpha_2, \alpha_3 \rangle \cap W_1 \cap W_2$ with probability $\{[(q^{d-1} - 1)/(q-1) - (q^{d-2} - 1)/(q-1)]/[(q^{d-1} - 1)/(q-1)]\} > 1 - 1/q$. It follows that, with probability $> (1 - 1/q)^4$, $J = \langle t_1, t_2, t_3 \rangle$ is irreducible on $U = \langle \alpha_1, \alpha_2, \alpha_3 \rangle$ and $V = U \oplus Z$, where $Z = W_1 \cap W_2 \cap W_3$.

Since each pair α_i, W_i determines the same number of transvections, this reduces considerations to the case of an irreducible subgroup J of $\mathrm{SL}(U)$ generated by 3 transvections. If J is not $\mathrm{SL}(3, q)$ then it is contained in one of the following: (1) $\mathrm{SL}(3, p^i)$ with $\mathrm{GF}(p^i) \subset \mathrm{GF}(q)$, (2) $\mathrm{SU}(3, q^{1/2})$ (or $\mathrm{SU}(3, 2)'$ when $q = 4$), (3) $(\mathbb{Z}_{q-1} \times \mathbb{Z}_{q-1}) \rtimes S_3$ when $p = 2$ or (4) $3A_6$ when $q = 4$ [Mi1, Ha]. Moreover, there is at most one G–conjugacy class of each of (2)-(3), one or three of type (1) (for any given i), and three of type (4). The probability that type (1) occurs is at most

$$\sum_{e/i \text{ is prime}} \frac{q^3(q^3-1)(q^2-1)}{p^{3i}(p^{3i}-1)(p^{2i}-1)} \frac{[(p^{3i}-1)(p^i+1)]^3}{[(q^3-1)(q+1)]^3} < 1/2q.$$

(Here, $|\mathrm{SL}(3,q)| = q^3(q^3-1)(q^2-1)$, and $|\mathrm{PSL}(3,q)| = |\mathrm{SL}(3,q)|/3$ when there are 3 conjugacy classes in cases (1) and (4). Also, $\mathrm{SL}(3,q)$ has exactly $(q^3-1)(q+1)$ nontrivial transvections.) The probability that (2), (3) or (4) occurs is estimated similarly. Hence, J is $\mathrm{SL}(3,q)$ with probability $> 1-1/q$. $\square$

This type of crude probability argument, similar to [KaLu], will be used for all generation estimates. We will also need similar information concerning subgroups of classical groups generated by other types of elements. The following is the simplest example of this:

Lemma 3.8 *Let $G = \mathrm{PSL}(2,q)$ with $q = p^e \geq 4$.*

(i) *Suppose that $J = \langle g, h \rangle$ is an irreducible subgroup of G with $|g| = |h| = \mathrm{ppd}^\#(p; e)$ or $\mathrm{ppd}^\#(p; 2e)$. Then J is one of the following: G; cyclic; A_4, S_4 or A_5, where q is odd; or $\mathrm{PSL}(2, q^{1/2})$ or $\mathrm{PGL}(2, q^{1/2})$.*

(ii) *Suppose that either $q \geq 8$ is even, or that $q \geq 17$. Then, for $i = 1$ or 2, two elements of the same $\mathrm{ppd}^\#(p; ie)$-order generate G with probability > 0.55. When $q = 4$ this probability is ≥ 0.3.*

(iii) *If $q \neq 9$ then, with probability $\geq 1/4$, two nontrivial transvections in G generate*

$$\begin{cases} G & \text{if } p \neq 2 \\ \text{a dihedral group of order } 2\mathrm{ppd}^\#(2, 2e) & \text{if } p = 2. \end{cases}$$

For all q, with probability $\geq 1/4$ three nontrivial transvections generate G.

Proof. (i) The list of subgroups of G has been known for almost a century [Mo, Wi] (cf. the more familiar reference [Di, pp. 285-6]), and leads to the first statement. Moreover, there are at most two conjugacy classes of each isomorphism type of subgroups A_4, S_4 or A_5, and at most one of $\mathrm{PGL}(2, q^{1/2})$.

(ii) The case $G = \mathrm{PSL}(2,4) \cong A_5$ is very elementary, so assume that $q \geq 8$. We will estimate the probability that $J := \langle g, h \rangle < G$, where now

J might be reducible. This reducible case occurs with probability at most $(q+1) \cdot q^2/\{q(q+1)/2\}^2 = 4/(q+1)$. Also, J lies in $\mathrm{PGL}(2, q^{1/2})$ with probability at most $|\mathrm{PSL}(2,q)\colon \mathrm{PGL}(2,q^{1/2})|\, \{(q^{1/2}(q^{1/2}-1))\}^2/\{q(q+1)\}^2 = (q^{1/2}-1)^2/(2,q-1)q^{1/2}(q+1)$. Next, J lies in an irreducible cyclic group with probability $1/\{(2,q)(q+1)\}$. Adding probabilities completes the proof when q is even, so assume that q is odd.

The only other cases to consider are when $|g| = |h|$ is 3, 4 or 5. We must take into account the fact that there can be two conjugacy classes of isomorphic subgroups J. It is easy to check that the cases $|g| = |h| = 3$, $J \cong A_4$; $|g| = |h| = 3$, $J \cong A_5$; $|g| = |h| = 4$, $J \cong S_4$; and $|g| = |h| = 5$, $J \cong A_5$, occur with probability at most $c(q^2-1)/q(q\pm1)^2$, where c is 2, 1, 2, 8/3 in the respective cases. Since $q \geq 17$, this yields the desired estimate.

(iii) This is proved as in (ii). (N.B.—If $q = 9$ then $G \cong A_6$, and two conjugate elements of order 3 never generate G.) □

Corollary 3.9 *Let $G = A \times B$ with $A \cong B \cong \mathrm{PSL}(2,q)$. If $q \geq 8$ is even or if $q \geq 17$ then, for $i = 1$ or 2, two elements of G, each projecting onto elements of A and B of the same $\mathrm{ppd}^\#(p;ie)$–order, generate G with probability $\geq 1/20$.*

If $q = 4$ then two such elements of order 3 generate at least A with probability $\geq 1/6$.

Proof. Assume that $q \geq 8$ is even or that $q \geq 17$. For $j = 1, 2$ we are considering elements $g_j = a_j b_j$, $a_j \in A$, $b_j \in B$. By the lemma, with probability $> 1 - 0.45 - 0.45$ we have both $\langle a_1, a_2 \rangle = A$ and $\langle b_1, b_2 \rangle = B$. Then $J := \langle (a_1 b_1, a_2 b_2) \rangle$ is G unless $J = \{(a, a^x) \mid a \in A\}$ for some $x \in \mathrm{Aut}A$ (where we are identifying A and B). In this event $b_1 = a_1^x$, and we need to know the probability that we also have $b_2 = a_2^x$. Given a_1, b_1, there are $(2, q-1)(q\pm1)e$ choices for x (corresponding to $\mathrm{N}_{\mathrm{Aut}A}(\langle a_1 \rangle)$). Thus, the probability that b_2 also satisfies $b_2 = a_2^x$ is $(2, q-1)(q\pm1)e/q(q\mp1) < 1/2$, so that $J = G$ with probability $> 0.1 \cdot 1/2$.

Similarly, if $q = 4$ then the desired probability is at least $1/6$. □

3.2   Transvection groups

We now begin our algorithm when $G \cong \mathrm{PSL}(d,q)$ or $\mathrm{SL}(d,q)$. We *assume that $d \geq 4$*: the cases $d = 2$ and $d = 3$ can be dealt with by brute force, but will be handled within our time constraints in **3.6**.

For timing we always use the quantities ξ, μ appearing in Theorem 1.1.

Sections **3.2–3.4** and **3.5.1** can be viewed as preprocessing, while **3.5.2** and **3.5.3** are the main parts of the algorithms.

In Sections 3–6 there will always be a vector space V underlying the classical group G; we will seek to construct this space and, especially, the projective action of G on it. Until it has been constructed, we can use the existence of V for explanations and correctness proofs, but not for algorithmic purposes.

3.2.1 Finding transvections

We separate the cases $d > 4$ and $d = 4$. Let $z := q^{d-2} - 1$.

(1) **Case** $d > 4$: Choose up to $\lceil 4q(d-2)\ln(2d)\rceil$ elements $\tau \in G$ to find one of $p \cdot \mathrm{ppd}^{\#}(p; e(d-2))$–order. If no such element is found, report failure. Otherwise, we claim that $t := \tau^z$ *is a transvection.*

Namely, τ^p has one irreducible constituent of degree $d-2$, the remaining ones spanning a 2–space of the target vector space V. Then τ^z induces 1 on the first of these and a transvection on the second, and hence is a nontrivial transvection of V. We will use this type of argument repeatedly in the future, usually without further comments.

(2) **Case** $d = 4$: We may assume that $q > 17$ if $q - 1$ is a power of 2, as otherwise brute force can be used to identify G. Choose up to $96q$ elements to find $\tau \in G$ of $p \cdot \mathrm{ppd}^{\#}(p; 2e) \cdot \mathrm{ppd}^{\#}(p; e)$–order (this means that $|\tau|$ is divisible by the integers involved in the definition of these $\mathrm{ppd}^{\#}$; cf. **2.4**); we also require that $\tau^{8p(q+1)} \neq 1$ if $q - 1$ is a power of 2. Once again we claim that $t := \tau^z$ *is a transvection.* As in (1), this is the case if τ leaves invariant a decomposition of V into the direct sum of two 2–spaces, since then $\tau^{p(q-1)}$ centralizes some 2–space.

In order to see that there is no other possibility we again use the element $\tau^{p(q-1)}$ of order dividing $q + 1$. If it does not act homogeneously on V then we have the situation of the preceding paragraph. If it acts homogeneously on V, then the centralizer in $\mathrm{GL}(4, q)$ of the corresponding linear transformation is $\mathrm{GL}(2, q^2) \cdot 2$. Then $\mathrm{C}_G(\tau^{p(q-1)})$ is a homomorphic image of $(\mathrm{GL}(2, q^2) \cdot 2) \cap \mathrm{SL}(4, q)$; but for an element of order p in that group the order of its centralizer is a factor of $|\mathrm{Z}(G)|2q(q + 1)$, where $|\mathrm{Z}(G)| \big| (4, q - 1)$. This contradicts our requirements concerning $|\tau|$.

Time: Las Vegas $O(qd \log d[\xi + \mu ed^2 \log q])$ in either case, since for each of the $O(qd \log d)$ elements tested it takes time $O(\mu(ed)^2 \log p)$ to find τ^z and to check the ppd property using Lemma 2.6.

Reliability: $\geq 1 - 1/4d^2$.

For, $C_G(\tau)$ preserves a decomposition $V = V_2 \oplus V_{d-2}$ with $\dim V_i = i$. Then $C_G(\tau) = A \times B$ for a transvection group A of order q and an abelian group B of order dividing $q^{d-2} - 1$ (since determinants must be 1). There are $|G|/q(q-1)|B|(d-2)$ conjugates $(AB)^g$, $g \in G$, of AB in G (cf. **2.4**). If $d > 4$ then each $(AB)^g$ has at least $(q-1)|B|/2$ elements behaving in the manner we required for τ, and for each such element the centralizer is just $(AB)^g$. Thus, the desired probability is at least $\{1/|B|q(q-1)(d-2)\} \cdot \{(q-1)|B|/2\} = 1/2q(d-2)$ (cf. Lemma 2.5). Hence, the probability that none of our $\lceil 4q(d-2)\ln(2d) \rceil$ chosen elements has the desired order is at most $(1 - 1/2q(d-2))^{4q(d-2)\ln(2d)} < 1/4d^2$. The argument is similar if $d = 4$.

Once again, this type of argument often will be presumed.

3.2.2   Finding transvection groups and J

Choose up to $\lceil 2^6 \ln(8d^2) \rceil$ pairs t_1, t_2 of conjugates of t. By Lemma 3.7, for a single pair $J := \langle t, t_1, t_2 \rangle \cong \mathrm{SL}(3, q)$ with probability $\geq 1/2^5$. (Moreover, J then must split the target vector space V as $V = U \oplus Z$, where $\dim U = 3$, the center α of t is in U and the axis W of t contains Z.)

Use Corollary 3.20 (or see **3.2.3**) in order to test whether $J \cong \mathrm{SL}(3, q)$ and, if so, to obtain a set $\mathcal{S}_J^*$ of transvections generating J together with a vector space V_J underlying J, and an isomorphism $\lambda_J \colon J \to \mathrm{SL}(V_J)$. Use λ_J^{-1} to find the following:

(i) The transvection group T of J containing t;

(ii) The two subgroups X_1, X_2 of J having order q^2, containing T and consisting of transvections;

(iii) G_3, the subgroup of $\mathrm{N}_J(X_1)$ generated by its transvections, and a subgroup $D \cong \mathrm{SL}(2, q)$ of G_3 such that $G_3 = X_1 \rtimes D$ and $D\lambda_J$ fixes a nonincident point-hyperplane pair of V_J;

(iv) $j(\gamma) \in J$, a transvection such that D normalizes $X_2^{j(\gamma)}$ (we are thinking of X_1 as consisting of all transvections with the same axis in V_J as t and $X_2^{j(\gamma)}$ as consisting of all transvections of V_J whose center is a point γ of V_J not on this axis and fixed by D); and

(v) $j \in \mathrm{N}_J(T) \cap \mathrm{N}_J(X_2^{j(\gamma)})$ of order $q-1$ inducing on T an automorphism of order $q-1$, together with a straight-line program of length $O(\log q)$ from $\mathcal{S}_j^*$ to j.

The generating set $\mathcal{S}^*(3)$: In **3.3.3** we will define the generating set $\mathcal{S}^*$ in Theorem 1.1 as the union of three sets $\mathcal{S}^*(1)$, $\mathcal{S}^*(2)$ and $\mathcal{S}^*(3)$ of p–elements. The set $\mathcal{S}^*(1)$ will be obtained recursively in **3.3.3**, $\mathcal{S}^*(2)$ is defined in **3.3.1**, and $\mathcal{S}^*(3)$ is our set $\mathcal{S}_j^*$ of generators, increased so as to include sets of transvections generating T, X_1, X_2, and G_3, as well as $j(\gamma)$. We may assume that $|\mathcal{S}^*(3)|$ is $O(e)$.

Reliability: $> 1 - 1/8d^2$: by Lemma 3.7 a pair t_1, t_2 produces a group $J \cong \mathrm{SL}(3, q)$ with probability $\geq 1/2^5$; and if actually $J \cong \mathrm{SL}(3, q)$, then the call to Corollary 3.20 tests this isomorphism and outputs the desired isomorphism with probability $> 1/2$. Hence, a pair will produce an output with probability $> 1/2^6$. Repeating $\lceil 2^6 \ln(8d^2) \rceil$ times, we obtain an output with the stated probability.

Time: Las Vegas $O(\log d\{\xi q e + \mu q \log^2 q\})$ for $O(\log d)$ groups J in view of Corollary 3.20.

3.2.3 Isomorphism and nonisomorphism testing

Corollary 3.20, cited above, requires parts of **3.3–3.5**. Those sections could have been written as lemmas with long lists of hypotheses, which would have made them harder to understand. Instead we assume that a reader will readily verify the needed sections in the context of Corollary 3.20, although this makes it somewhat harder to check some aspects of our timing that are faster than mere brute force.

Nevertheless, brute force would indeed be a simpler approach to handle the groups J, despite the much poorer timing estimates. For example, construct the regular permutation representation of J in $O(\mu|J|^2)$ time, and then use a variety of possible procedures to find the underlying 3–dimensional vector space (such as those in [Ka3] and [KP]). Similar statements occur later as well. The following table contains all such brute force "simplifications", including sections into which the resulting inferior timings could be inserted. In each case J is constructed having at most 6 generators.

J	Section	Timing
$\mathrm{SL}(3,q)$	**3.2.2**	$O(\mu q^{16})$
$\mathrm{SL}(2,q) \circ \mathrm{SL}(2,q)$	**4.2.1**	$O(\mu q^{12})$
$\Omega^+(6,q)$	**5.2.2**	$O(\mu q^{30})$
$\Omega^-(8,q)$	**4.2.2**	$O(\mu q^{54})$
$\mathrm{Sp}(4,q)$	**5.2.1**	$O(\mu q^{20})$
$\mathrm{SU}(4,q)$	**6.2.2**	$O(\mu q^{30})$

In fact, this paper was originally written using just such shortcuts. The requirements of Theorem 1.7 forced us to find faster ways to handle small–dimensional groups.

In **3.2.2**, we also had to decide whether t, t_1, t_2 generated $\mathrm{SL}(3,q)$ or a proper subgroup of it. While brute force will nicely take care of the question of isomorphism and nonisomorphism of $\langle t, t_1, t_2 \rangle$ and $\mathrm{SL}(3,q)$, we also need *faster nonisomorphism testing*. Repetition of the algorithm for Theorem 1.1 or 1.1′ produces a Monte Carlo nonisomorphism test (this assumes that these are repetitions of the entire algorithm, including the generators and relations verification portion in Section 7). For the groups in the above list, a nonisomorphism test is possible in the contexts in which the groups occur: we construct a subgroup J generated by certain types of elements of G, and the number of possible subgroups of our classical group G that at most 6 of these elements can generate is very limited. In each case, only the desired group could survive the tests in **3.6.1**, **3.6.2**, **3.6.3**, **4.6.2**, **5.6.1**, **6.6.1** and **6.6.2**. This means that if the constructed group J is only a proper subgroup of the desired group, the algorithms in those sections will report failure.

3.3 Finding Q, $Q(\alpha)$, H, L and $\mathcal{S}^*$

3.3.1 Q, $Q(\alpha)$ and H

Let

$$Q := \langle X_1^{\tau^{p^i}} \mid 0 \le i < d-2 \rangle \quad \text{and} \quad Q(\alpha) := \langle X_2^{\tau^{p^i}} \mid 0 \le i < d-2 \rangle.$$

Then Q and $Q(\alpha)$ are generated by transvection groups, since X_1 and X_2 are. Let $\mathcal{S}^*(2)$ denote the union of the generating sets for the transvection groups $X_1^{\tau^{p^i}}$ and $X_2^{\tau^{p^i}}$, $0 \le i < d-2$; *we will store these subsets of $\mathcal{S}^*(2)$, and hence we will know, for each member of $\mathcal{S}^*(2)$, generators for the transvection group containing it.* Here $|\mathcal{S}^*(2)|$ is $O(ed)$ and $\mathcal{S}^*(2)$ contains transvections from $O(d)$ transvection groups.

Notational convention: We will use single and double quotation marks in this and later sections, according to the following convention: "Q" denotes the *matrix group* appearing in **3.1.3**; 'Q' denotes a copy of that group *lying in G and containing a subgroup Q already constructed.*

Interchanging the roles of V and V^* if necessary, we may assume that X_1 consists of transvections all having the same axis W, while X_2 consists of transvections all having the same center $\alpha \in W$ in the target space V. Then we may assume that Q lies in the group 'Q' consisting of all transvections having axis W. Since τ^p centralizes t, it leaves invariant W and α (cf. **3.1.3**). Then τ^p is irreducible on W/α, while X_1 contains T (cf. **3.2.2**(i)) as well as transvection subgroups with center not α. Thus, by Lemma 2.7, $Q = $ 'Q'. Similarly, $Q(\alpha)$ is the group of all transvections having center α. This proves the first two parts of the next result, which is fundamental to this section (as are analogues in Sections 4–6):

Lemma 3.10   (i) $Q = $ 'Q'.

(ii) $Q(\alpha)$ *is the group of all transvections having the same center as t.*

(iii) $H := \langle G_3^{\tau^{pi}} \mid 0 \le i \le d \rangle \cong \mathrm{ASL}(d-1,q) \cong Q \rtimes \mathrm{SL}(d-1,q)$.

Proof. Let W denote the hyperplane of V in **3.1.3** centralized by 'Q'. Write $W_1 = [W, G_3]$, so that $W_1 \cong X_1$, $\alpha \in W_1^{\tau^{pi}}$ and $W = W_1^{\tau^{pi}} \oplus \mathrm{C}_W(G_3^{\tau^{pi}})$ for each i. Every $G_3^{\tau^{pi}}$–invariant subspace of W either contains $W_1^{\tau^{pi}}$ or is contained in $\mathrm{C}_W(G_3^{\tau^{pi}})$.

For integers $a \le b$ let $H_{a,b} = \langle G_3^{\tau^{pi}} \mid a \le i \le b \rangle$, so that $H = H_{0,d}$.

We first show that $H_{0,d-3}$ is irreducible on W. If Y is an $H_{0,d-3}$–invariant subspace then, whenever $0 \le i \le d-3$, either $Y \supseteq W_1^{\tau^{pi}}$ or $Y \subseteq \mathrm{C}_W(G_3^{\tau^{pi}})$. Since $\alpha \in W_1^{\tau^{pi}}$ and $\alpha \notin \mathrm{C}_W(G_3^{\tau^{pi}})$, if $W_1^{\tau^{pi}} \subseteq Y$ for one i then it holds for all $0 \le i \le d-3$, so that Y contains the span of these subspaces $W_1^{\tau^{pi}}$, which is W by Lemma 2.7. Similarly, if $Y \subseteq \bigcap_0^{d-3} \langle \alpha, \mathrm{C}_W(G_3^{\tau^{pi}}) \rangle$ then $Y \subseteq \alpha$ by Lemma 2.7 applied to the dual of W/α, and hence $Y = 0$ since G_3 moves α.

It follows that $H_{0,d-3}/Q$ is an irreducible subgroup of $\mathrm{SL}(d-1,q)$ generated by transvection groups, so that Theorem 3.6 applies. If d is even then $H_{0,d-3}/Q$ is a subgroup of an odd–dimensional group $\mathrm{SL}(d-1,q)$, so that possibilities (ii-iv) cannot occur.

If $d = 2n+1$ is odd then we need to eliminate (ii-iv). Suppose that $H_{0,d}/Q < \mathrm{SL}(d-1,q)$. Note that τ^p cannot normalize any of the groups (ii-iv) since $|\tau^p| = \mathrm{ppd}^\#(p; e(2n-1))$. Thus, $H_{0,d-3} \ne H_{1,d-2}$, so that $H_{0,d-3} < H_{0,d-2}$.

Similarly, $H_{0,d-3} < H_{0,d-2} < H_{0,d-1} < H_{0,d} = H$. However, there is no strictly increasing sequence of four of the groups (ii-iv). This contradiction proves that $H_{0,d}/Q = \mathrm{SL}(d-1,q)$. $\square$

3.3.2   Effective transitivity of Q and $Q(\alpha)$; the complement L

Many times throughout this paper we will be dealing with an abstract group but using language stemming from the isomorphic concrete group of (projective) linear transformations. This dual view might cause trouble for a computer, but it is intended to help the reader (and authors) keep track of the underlying geometry.

Points: *Points* are defined to be the conjugates $Q(\alpha)^g$, $g \in G$. We write each point as $Q(\beta)$, just as a label, making it easier to regard α and β as points of the target vector space $V = \mathbb{F}^d$. Recall that we cannot list the entire set of points.

Equality testing: Two points are equal if and only if they commute; however, we will use a faster test. Let $1 \neq g \in Q(\alpha)$; then $Q(\alpha)^x = Q(\alpha)^y$ if and only if $[g^x, Q(\alpha)^y] = 1$, using one generator for each of the $O(d)$ transvection groups generating $Q(\alpha)^y$ (cf. **3.3.1**). (This condition implies that the center α^x of g^x is fixed by all members of $Q(\alpha)^y$.) **Time:** $O(\mu d)$.

Incidence: We are thinking of Q as identified with the hyperplane W (of V) it will eventually fix pointwise. With this in mind, in view of **3.1.2**(iii,iv), for $x \in G$ we say that $Q(\beta)$ is *on* Q^x (or Q^x is *on* $Q(\beta)$) if and only if $[[Q(\beta), Q^x], Q^x] = 1$ (in which case $[[Q(\beta), Q^x], Q(\beta)] = 1$). Once again we will use a faster test: it is only necessary to test whether $[[u,v],v] = 1$ for one $u \neq 1$ in $Q(\beta)$ while letting v range over a single generator of each transvection group from among the generators of Q^x (cf. **3.3.1**). **Time:** $O(\mu d)$ for a given pair $Q^x, Q(\beta)$.

Lemma 3.11   (i) *In deterministic $O(\mu[qe + qd])$ time, given two points $Q(\beta)$, $Q(\delta)$ not on Q, an element of Q can be found conjugating $Q(\delta)$ to $Q(\beta)$.*

(ii) *In deterministic $O(\mu[qe + qd])$ time, given two conjugates Q^x, Q^y not on $Q(\alpha)$, an element of $Q(\alpha)$ can be found conjugating Q^x to Q^y.*

Proof. By symmetry we only need to consider (i). Let a_0 be a generator of $Q(\beta)$, let $A < Q(\beta)$ be the transvection group containing it (cf. **3.3.1**), and list A.

If $Q(\delta)^{a_0} \neq Q(\delta)$, let b be any generator of $Q(\delta)$, find $a \in A$ such that $Q^{b^a} = Q$, and let $y := b^a$.

If $Q(\delta)^{a_0} = Q(\delta)$, find a generator b of $Q(\delta)$ such that $c := [a_0, b] \neq 1$, find $a \in A$ such that $Q^{ca} = Q$, and let $y := ca$.

Find a generator x of Q such that $[x, y] \neq 1$; if this does not exist then $Q(\delta)^{a_0} \neq Q(\delta)$, and we replace b by any other generator of $Q(\delta)$ not lying in the same transvection group as b. Find the transvection group X in Q containing x. Find and list $[X, y]$. Find $u \in [X, y]$ such that $Q(\beta)^u = Q(\delta)$. Output u.

Correctness: Let W denote the axis of Q, and write $\varepsilon = \langle \beta, \delta \rangle \cap W$.

If $Q(\delta)^{a_0} \neq Q(\delta)$ then A is transitive on the points of $\langle \beta, \delta \rangle - \{\beta\}$, so that there exists $a \in A$ such that the center of b^a is on Q; that is, $\delta^a = \varepsilon$. This occurs for a if and only if $Q^{b^a} = Q$. It may happen that $[b^a, Q] = 1$, which occurs only for b in the transvection group with center δ and axis containing the intersection of those of Q and A; hence an additional choice of b avoids this possibility. (Note that a does not have to be recomputed in this event.) By **3.1.2**, $[b^a, X] = T(\varepsilon, W)$, which contains the desired element u.

If $Q(\delta)^{a_0} = Q(\delta)$ then A fixes $Q(\delta)$ but does not centralize $Q(\delta)$. Then b exists, and c has center δ and the same axis W' as A by **3.1.2**. Both A and c act on the set of hyperplanes $\neq W'$ containing $W \cap W'$, neither has center in W, and A acts transitively on this set of hyperplanes. Thus, a exists.

Since $\langle A, c \rangle$ consists of transvections with axis W' and centers in $\langle \beta, \delta \rangle$, ca must have center ε and axis W'. By **3.1.2**, $[ca, X] = T(\varepsilon, W)$, which once again contains the desired element u.

Time: $O(\mu(qe + qd))$. Listing of elementary abelian groups of order q takes $O(\mu q e)$ time using Lemma 2.1. We just saw that equality testing can be accomplished in $O(\mu d)$ time. If $Q(\delta)^{a_0} \neq Q(\delta)$, find a in $O(q \cdot \mu d)$ time (in view of our remark concerning incidence testing), x in $O(\mu d)$ time (testing only one generator per transvection group), $[b^a, X]$ in $O(\mu q e)$ time, and u in $O(q \cdot \mu d)$ time. If $Q(\delta)^{a_0} = Q(\delta)$, find b in $O(\mu d)$ time, a in $O(q \cdot \mu d)$ time, $[ca, X]$ in $O(\mu q e)$ time, and u in $O(q \cdot \mu d)$ time. $\square$

Remark. By (3.2), the element of Q found in the lemma is *unique*, and there is a complement to Q in G'_α. One of these complements fixes the point $Q(\gamma) := Q(\alpha)^{j(\gamma)}$ (cf. **3.2.2**(iv)).

Corollary 3.12 *In deterministic* $O(\mu[qed + qd^2])$ *time a subgroup L can be found such that $H = Q \rtimes L$, where L fixes the point $Q(\gamma)$.*

Proof. We have $H = Q\langle D^{\tau^{p^i}} \mid 0 \leq i \leq d\rangle$, where $D \cong \mathrm{SL}(2, q)$ in **3.2.2**(iii) fixes $Q(\gamma)$, which is moved by X_1 and hence by Q. For $0 \leq i \leq d$ use the preceding lemma to find $u_i \in Q$ such that $Q(\gamma)^{\tau^{p^i} u_i} = Q(\gamma)$. Let $L := \langle D^{\tau^{p^i} u_i} \mid 0 \leq i \leq d\rangle$. Since $D^{\tau^{p^i} u_i} Q = D^{\tau^{p^i}} Q$ and $D^{\tau^{p^i} u_i}$ fixes $Q(\gamma)$, we have $H = QL$ and $Q \cap L = 1$, and hence $H = Q \rtimes L$.

This requires $O(d)$ uses of the algorithm in the preceding lemma. □

Remarks. The group $L^{j(\gamma)^{-1}}$ is a complement to $Q(\alpha)$ in G'_α. Thus, we have complete symmetry between Q and $Q(\alpha)$: anything we do for Q and H can also be done for $Q(\alpha)$ and G'_α.

At this point we have the group $L \cong \mathrm{SL}(d-1, q)$ and its $d-1$–dimensional module Q. What we lack is a matrix produced by any given element of L (cf. **3.4.3**(5)).

3.3.3 Recursion: $\mathcal{S}^*$ and λ_L

We will require that *our set $\mathcal{S}^*$ of generators of G consists entirely of transvections.* This property will be preserved throughout our recursive calls in this section; there are analogues in Sections 4–6 using "root elements" in place of transvections. The restriction to elements of order p is needed for Theorem 1.1′, and hence for the recursive call used for Theorem 1.1.

The set $\mathcal{S}^*$: Our group $G = \langle H, J\rangle$ is generated by the set

$$\mathcal{S}^* := \mathcal{S}^*(1) \cup \mathcal{S}^*(2) \cup \mathcal{S}^*(3),$$

where $\mathcal{S}^*(1) = \mathcal{S}_L^*$ will be obtained below recursively while $\mathcal{S}^*(2)$ and $\mathcal{S}^*(3)$ were defined in **3.3.1** and **3.2.2**. Then $|\mathcal{S}^*(2)|$ is $O(ed)$ and $|\mathcal{S}^*(3)|$ is $O(e)$, so $|\mathcal{S}^*|$ is $O(ed^2)$.

The choice of λ_L: Recursively use Theorem 1.1′ to obtain *a $d-1$–dimensional* $\mathbb{F}$–*space V_L, a set $\mathcal{S}_L^*$ of generators of L and an isomorphism* $\lambda_L \colon L \to \mathrm{SL}(V_L) = \mathrm{SL}(d-1, q)$ *defined on* $\mathcal{S}_L^*$.

We may need to replace λ_L by $\lambda_L \iota$, where ι denotes the inverse transpose map. Both λ_L and $\lambda_L \iota$ produce isomorphisms $L \to \mathrm{SL}(d-1, q)$, but only one of them will give the action needed: we need to know that, when we fix a vector using $L\lambda_L$, the preimage will also fix a nontrivial element of Q. We achieve this as follows.

In the derived subgroup P_0 of the stabilizer in $L\lambda_L$ of a 1-space find the normal subgroup Q_0 of order q^{d-2} (cf. **3.1.3**). Test whether $[[Q, Q_0\lambda_L^{-1}], P_0\lambda_L^{-1}] \neq 1$; if so then replace λ_L by $\lambda_L \iota$.

Correctness: If $[Q, Q_0 \lambda_L^{-1}]$ is a hyperplane of Q then $P_0 \lambda_L^{-1}$ acts nontrivially on it; but if $[Q, Q_0 \lambda_L^{-1}]$ is a 1–space of Q then $P_0 \lambda_L^{-1}$ centralizes it.

Time: $O(\mu d^3 e \log q)$, using Proposition 3.17 to apply λ_L^{-1}; we may assume that P_0 has $O(1)$ generators. (N.B.—This speed can be improved to $O(\mu d^2 \log q)$ by replacing Q_0 by the group of order p^2 generated by two of its elements lying in different transvection groups.)

Now λ_L has the property that *it sends the stabilizer of some nontrivial element of Q to the stabilizer of some nontrivial element of "Q"*.

Remark concerning timing and the recursive call: When we make a recursive call we will refer to **3.5.3, 3.6.4**, or their analogues for other classical groups, rather than to Theorems 1.1 or 1.1$'$, since those sections contain more precise timing estimates than the theorems.

Computing preimages recursively: We have just applied λ_L^{-1} to elements of $L\lambda_L$, and this will be done often. This is the content of Proposition 3.17 (and of Theorem 1.1$'$(iv)), with L in place of G: *in $O(\mu d^2 \log q)$ time $h\lambda_L^{-1}$ can be found for any given $h \in L\lambda$*. This result could be proved now, and for timing purposes perhaps should be proved now. However, the pattern of its proof is needed for the closely related Proposition 3.18, which cannot be proved until we have more machinery available.

The same observation will also be used later in **4.3.3, 5.3.3** and **6.3.3**.

We also note that, when using Proposition 3.17, we obtained a timing of $O(\mu d^2 \log q)$ instead of $O(\mu d^2 \log q + d^4 \log q)$. For, $d^4 \log q$ is $O(\mu d^2 \log q)$ since we have assumed that $\mu \geq N \geq (d^2 \log q)/2$ (cf. **2.2**). We will use this observation repeatedly without further comment.

The isomorphism $\lambda^\#$: Throughout this section we are assuming that there exists an isomorphism $\lambda^\#: G \to \mathrm{PSL}(V)$ or $\mathrm{SL}(V)$. While the goal is to construct such an isomorphism, we can use its existence to motivate and prove correctness of subroutines in the algorithm. If necessary, replace V by its dual space in order to assume that $\lambda^\#$ sends the nonincident point–hyperplane pair $Q(\gamma), Q$ to a nonincident point–hyperplane pair of V; we may assume that this pair is γ, W in the notation of **3.1.3**. Recall the convention in **3.3.1**; for example "Q" denotes the matrix group studied in **3.1.3**. Then $\lambda^\#$ sends $Q \to$ "Q" and $L \to$ "L"$:= L\lambda_L = \mathrm{SL}(d-1, q)$. Note that $\lambda_L \lambda^{\#-1}$ induces an automorphism of L sending the stabilizer of some 1–space of Q to the stabilizer of some 1–space of Q (we just arranged this). It follows that $\lambda_L \lambda^{\#-1}$ is induced (via conjugation) by a semilinear map $g: Q \to Q$, and hence that λ_L

extends to the isomorphism $\lambda^\bullet \colon H = QL \to$ "Q" "L" given by $h \mapsto h^g \lambda^\#$. By following $\lambda^\#$ with a field automorphism we see that there is an isomorphism $\lambda^\bullet$ of the desired sort such that the map $g = \lambda^\bullet|_Q$ is linear. In Lemma 3.14 we will find $\lambda^\bullet|_Q$. For now we note only that $\lambda^\bullet|_Q$ is uniquely determined up to a scalar transformation.

3.4 Deterministic algorithms for Q

All procedures in 3.4 and 3.5 will be deterministic.

3.4.1 Decomposition of Q and $Q(\alpha)$

Lemma 3.13 *In deterministic $O(\mu d^2 \log q)$ time $O(e)$–generator subgroups $A_i \le Q$, $1 \le i \le d-1$, can be found such that the following hold:*

(i) *$|A_i| = q$ and $Q = A_1 \times \cdots \times A_{d-1}$; and*

(ii) *In deterministic $O(\mu d)$ time, any given $u \in Q$ can be expressed in the form $u = \prod_1^{d-1} a_i$ with $a_i \in A_i$.*

The same result holds for $Q(\alpha)$ in place of Q.

Proof. We use the fact that Theorem 1.1$'$(iv) holds for L recursively. We have an L–module $V_L = \mathbb{F}^{d-1}$ (cf. **3.3.3**, **2.3**) with standard basis $e_1, \ldots, e_{d-1}$. Find $c' := (e_1 \mapsto \cdots \mapsto e_{d-1} \mapsto (-1)^{d-2} e_1) \in \mathrm{SL}(d-1, q)$. Let E_{ij} be the $d-1 \times d-1$ matrix over $\mathbb{F}$ having (i, j) entry 1 and all other entries 0. Find $c := c'\lambda_L^{-1}$, $t_{21} := (I + E_{21})\lambda_L^{-1}$ and $t_{j+1,j} := t_{21}^{c^{j-1}}$ for $2 \le j \le d-2$.

Find a generator $r \in Q$ such that $[r, t_{21}] \ne 1$, and let $R < Q$ be the transvection group containing it (cf. **3.3.1**). Then $[Q, t_{21}] = [R, t_{21}]$ has $O(e)$ generators. Let

$$A_1 := [Q, t_{21}] \quad \text{and} \quad A_i := A_1^{c^{i-1}}, \ 2 \le i \le d-1.$$

Then Q is the direct product of the subgroups A_i. Given any $u \in Q$, we have $u = \prod_1^{d-1} a_i$, $a_i \in A_i$, where $a_i := [u, t_{i,i-1}]^c$ for $2 \le i \le d-1$ and $a_1 := u(\prod_2^{d-1} a_i)^{-1}$.

Correctness: If $v = \sum_i k_i e_i \in V_L$ then $k_i e_i = [v, I + E_{i,i-1}]^{c'}$ for $2 \le i \le d-1$ by elementary calculations. At the end of **3.3.3** we saw that there is an isomorphism $\lambda^\bullet \colon QL \to$ "Q" "L" extending λ_L. Then, using (3.4) and the isomorphism in (3.3), we have $r(v)\lambda^{\bullet-1} = \prod_i r(k_i e_i)\lambda^{\bullet-1}$ with $r(k_i e_i)\lambda^{\bullet-1} =$

$[r(v)\lambda^{\bullet -1}, t_{i,i-1}]^c \in [r(V_L), I + E_{i,i-1}]^{c'}\lambda^{\bullet -1} = A_i$ for $2 \le i \le d-1$, as asserted for an arbitrary element $u = r(v)\lambda^{\bullet -1}$ of Q.

Time: It takes $O(d^2)$ time to write the two required matrices c' and $I + E_{21}$, $O(\mu d^2 \log q)$ to apply λ_L^{-1} to them (using Proposition 3.17 with L in place of G), $O(\mu d)$ to find r, and $O(\mu e d)$ to conjugate A_1 and t_{21} by the required powers of c.

In order to handle $Q(\alpha)$ in the same manner, recall that L normalizes $Q(\gamma) = Q(\alpha)^{j(\gamma)}$ (cf. **3.3.2**). Thus, it suffices to use $t'_{12} := I - E_{12}$ (the inverse transpose of t'_{21}), and c as above in order to obtain $Q(\alpha) = Q(\gamma)^{j(\gamma)^{-1}} = \prod_{i=0}^{d-2}[Q(\gamma), t'_{12}\lambda_L^{-1}]^{c^{i-1}j(\gamma)^{-1}}$. $\square$

3.4.2   A field of endomorphisms of Q

We have a field $\mathbb{F}$ and generator ρ in **2.3**. Let $s'_1 := \begin{pmatrix} \rho & 0 & O \\ 0 & \rho^{-1} & O \\ O & O & I \end{pmatrix} \in$ SL$(d-1, q)$. Use Proposition 3.17 to find $s_1 := s'_1\lambda_L^{-1}$ and $s_i := s_1^{c^{i-1}}$ for $2 \le i \le d-1$ (where c was defined in **3.4.1**). Now define an automorphism s of $Q = A_1 \times \cdots \times A_{d-1}$ by

$$s := \text{conjugation by } s_i \text{ on } A_i, \ 1 \le i \le d-1.$$

Then GF$(q) := \langle s \rangle \cup \{0\}$ is the centralizer of the group of automorphisms of Q induced by L using conjugation.

Time: $O(\mu d^2 \log q)$ to find all s_i and s.

Identification of GF(q) and $\mathbb{F}$: We identify this field with $\mathbb{F}$ using $s \mapsto \rho$.

3.4.3 Q and $Q(\alpha)$ as $\mathbb{F}$–spaces and GF(p)–spaces

We can now compute with Q both as an $\mathbb{F}$–space and a GF(p)–space.

(1) *Find and compute with an $\mathbb{F}$–basis $\mathcal{B}$ and a GF(p)–basis $\mathcal{B}_p$ of Q*: We use the notation of **3.4.1** and **3.4.2**. Fix $1 \ne b_1 \in A_1$, and let $b_i := b_1^{c^{i-1}}$ for $1 \le i \le d-1$. Then $\mathcal{B} := \{b_i | 1 \le i \le d-1\}$ is an $\mathbb{F}$–basis and $\mathcal{B}_p := \{b_i^{s^j_i} | 1 \le i \le d-1, 0 \le j < e\}$ is a GF(p)–basis of Q.

For $1 \le i \le d-1$, we list A_i as $\{1\} \cup \{b_i^{s^j_i} | 0 \le j < q-1\}$ in the notation of **3.4.2**.

Recall that in **2.3**, we stored the coefficients of the linear combination in the GF(p)–basis $\{1, \rho, \ldots, \rho^{e-1}\}$ for each element of $\mathbb{F}$. These coefficients can be used to express all elements of A_i as linear combinations of $A_i \cap \mathcal{B}_p$. **Time:** $O(\mu q d)$.

We now handle $Q(\alpha)$ similarly. We have generators for $Q(\gamma) = Q(\alpha)^{j(\gamma)}$. The end of the proof of Lemma 3.13 uses the decomposition $Q(\gamma) = A_1'' \times \cdots \times A_{d-1}''$, where $A_1'' := [Q(\gamma), t_{12}' \lambda^{-1}]$ is a transvection group in $Q(\gamma)$ and $A_i'' := (A_1'')^{c^{i-1}}$ for $2 \leq i \leq d-1$. Pick $1 \neq b_1'' \in A_1''$ and let $b_i'' := (b_1'')^{c^{i-1}}$. Then $\mathcal{B}'' := \{b_i'' \mid 1 \leq i \leq d-1\}$ is an $\mathbb{F}$–basis of $Q(\gamma)$ and $\mathcal{B}_p'' := \{(b_i'')^{s_i^{-j}} \mid 1 \leq i \leq d-1, 0 \leq j < e\}$ is a GF(p)–basis of $Q(\gamma)$. Finally, $\mathcal{B}' := (\mathcal{B}'')^{j(\gamma)^{-1}}$ is an $\mathbb{F}$–basis and $\mathcal{B}_p' := (\mathcal{B}_p'')^{j(\gamma)^{-1}}$ is a GF(p)–basis of $Q(\alpha)$.

(2) *Modify $\mathcal{S}^*(2)$*: We may assume that $\mathcal{S}^*(2)$ contains $\mathcal{B}_p$ and $\mathcal{B}_p'$.

Note that, if a transvection group contains a member of $\mathcal{B}_p$ or $\mathcal{B}_p'$, then generators for the transvection group are inside $\mathcal{S}^*(2)$. We assume that $\mathcal{S}^*(2)$ has been stored as a union of subsets, each subset being the intersection of $\mathcal{S}^*(2)$ with a transvection group (cf. **3.3.1**).

(3) *Computing the action of a given $\rho^j \in \mathbb{F}$ on a given $u \in Q$*: write $u = \prod_i u_i$ with $u_i \in A_i$ using Lemma 3.13(ii); and then $u\rho^j = \prod_i u_i^{s_i^j}$. **Time:** $O(\mu d \log q)$.

(4) *Linear combinations*: Any $u \in Q$ can be written as an $\mathbb{F}$–linear combination of $\mathcal{B}$, as follows. Use Lemma 3.13 to write $u = \prod_i u_i$ with $u_i \in A_i$; looking up u_i in the listing of A_i, we can find the coefficient of b_i in the $\mathbb{F}$–linear combination of each u_i in the basis $\mathcal{B}$. **Time:** $O(\mu q d)$.

Straight-line programs: Given a "linear combination" $u = \prod_{i \in I} b_i^{s_i^{j(i)}}$ of some $u \in Q$, where $I \subseteq \{1, \ldots, d-1\}$, by looking up the coefficients of $\rho^{j(i)}$ in our GF(p)–basis of $\mathbb{F}$ we obtain a GF(p)-linear combination of u in the basis $\mathcal{B}_p$. Hence we can also write a straight-line program of length $O(d \log q)$ reaching u from $\mathcal{B}_p$. **Time:** $O(\mu d \log q)$.

(5) *Matrices $\tilde{g}$*: Using $\mathcal{B}$ and (4) we can find the $d-1 \times d-1$ matrix $\tilde{g}$ representing the linear transformation induced on the $\mathbb{F}$–space Q by any given $g \in N_G(Q)$. **Time:** $O(d \cdot \mu q d)$ by (4).

3.4.4   The extension $\lambda^\bullet$

We have two isomorphisms of L with SL$(d-1, q)$: one of them, λ_L, coming from the recursive call, and another one, $h \mapsto \tilde{h}$, obtained in **3.4.3**(5) from the action of L on Q. We now show that these two isomorphisms coincide, and extend λ_L to all of H. Recall that, in Lemma 3.13 and **3.4.3**, we used the standard basis $e_1, \ldots, e_{d-1}$ of $V_L = \mathbb{F}^{d-1}$ in order to find a basis $b_1, \ldots, b_{d-1}$ of Q. Identify V_L with the $d-1$–space $W \subset \mathbb{F}^d$ occurring in **3.1.3** by appending a last coordinate 0 to each vector in $\mathbb{F}^{d-1}$. Recall the isomorphism $r : W \to$ "Q" in (3.3).

Lemma 3.14 λ_L can be extended to an isomorphism $\lambda^\bullet : H = QL \to$ "Q" "L" $\cong \mathbb{F}^{d-1} \rtimes \mathrm{SL}(d-1, q) = \mathrm{ASL}(d-1, q)$ with the following properties:

(i) if $b \in Q$ with $b = \sum_i k_i b_i$ (for $k_i \in \mathbb{F}$), then $b\lambda^\bullet = r(\sum_i k_i e_i)$, so that $\lambda^\bullet$ is $\mathbb{F}$–linear on Q; and

(ii) $\tilde{l} = l\lambda^\bullet = l\lambda_L$ whenever $l \in L$.

Proof. At the end of **3.3.3** we saw that λ_L extends to an isomorphism $\lambda^\bullet : QL \to$ "Q" "L" induced by a semilinear map $\lambda^\bullet|_Q : Q \to$ "Q" $= r(W) = r(V_L)$ that is uniquely determined up to a scalar transformation. We claim that the definition

$$b_i \lambda^\bullet := r(e_i) \text{ for } 1 \leq i \leq d-1$$

is forced by the known actions of L and $L\lambda_L$ (up to the aforementioned scalar ambiguity). Since $t_{21} = (I + E_{21})\lambda_L^{-1}$ we must have $\langle b_1 \lambda^\bullet \rangle = [Q, t_{21}]\lambda^\bullet = [r(W), I + E_{21}] = \langle r(e_1) \rangle$ (cf. (3.4)). Thus, $b_1 \lambda^\bullet$ must be in $\langle r(e_1) \rangle$, and we may assume that $b_1 \lambda^\bullet = r(e_1)$. By **3.4.1** we have $c : b_1 \mapsto \cdots \mapsto b_{d-1}$ and $c\lambda_L : e_1 \mapsto \cdots \mapsto e_{d-1}$. Using (3.4) we see inductively that $b_i \lambda^\bullet = b_1^{c^{i-1}} \lambda^\bullet = r(e_{i-1})^{c\lambda^\bullet} = r(e_i)$ for $2 \leq i \leq d-1$, as claimed.

Also, $\lambda^\bullet|_Q$ must be semilinear, but in fact it is linear: by the definition in **3.4.2**, s_1' induces ρ on $\langle r(e_1) \rangle$ while s and s_1 induce ρ on A_1, so that linearity follows from $(\rho b_1)\lambda^\bullet = (b_1^{s_1})\lambda^\bullet = (b_1 \lambda^\bullet)^{s_1 \lambda^\bullet} = (b_1 \lambda^\bullet)^{s_1'} = \rho(b_1 \lambda^\bullet)$. Now (i) is clear.

(ii) Write $\tilde{l} = (m_{ij})$, where $b_i^l = \sum_j m_{ij} b_j$ with $m_{ij} \in \mathbb{F}$ using the definition in **3.4.3**(5). Apply $\lambda^\bullet$ and use linearity to obtain $e_i^{l\lambda^\bullet} = \sum_j m_{ij} e_j$, so that $l\lambda^\bullet = (m_{ij})$. $\square$

Remarks. The spaces Q and V_L are related via $b = \sum_i k_i b_i \mapsto \dot{b} = \sum_i k_i e_i$, so that

$$b\lambda^\bullet = r(\dot{b}) \text{ for all } b \in Q. \tag{3.15}$$

We defined $\lambda^\bullet$ using its restrictions to L and Q. While we will not need to apply $\lambda^\bullet$ to an arbitrary $h \in H$, this would be easy to do since $h = ul$

with $u \in Q$ and $l \in L$, where u satisfies $Q(\gamma)^{hu^{-1}} = Q(\gamma)$ and is found using Lemma 3.11.

3.5   The homomorphism λ and straight-line programs

We have now finally set up the data structures needed for the proof of Theorems 1.1 and 1.1′ for the present G. Everything done so far is essentially just preprocessing for that proof.

Points were defined in **3.3.2**.

3.5.1   Labeling points

We now label any given *point by a 1–space in $V := \mathbb{F}^d$ in an H–invariant manner.*

Recall that $Q(\gamma)$ is not on Q. Label it by the 1–space $\langle 0, \ldots, 0, 1 \rangle$ of $\mathbb{F}^d$.

Given $Q(\delta) \neq Q(\gamma)$ not on Q, use Lemma 3.11(i) to find the unique $b \in Q$ conjugating $Q(\gamma)$ to $Q(\delta)$, find $b\lambda^{\bullet} \in$ "Q" using **3.4.3**(4) and Lemma 3.14(i) and (3.15), and label $Q(\delta)$ as $\langle 0, \ldots, 0, 1 \rangle (b\lambda^{\bullet}) = \langle \dot{b}, 1 \rangle$ using (3.15) and (3.3).

Finally, we need to label any given point $Q(\beta)$ on Q. Test the generators a of $Q(\beta)$ in order to find one such that $b := [a, b_i] \neq 1$ for $i = 1$ or 2. Then $b \in Q$, and we label $Q(\beta)$ by $\langle (0, \ldots, 0, 1)(b\lambda^{\bullet}) - (0, \ldots, 0, 1) \rangle = \langle \dot{b}, 0 \rangle$. (See (3.3) and (3.15). Note that we used the fact that the elements b_1 and b_2 in **3.4.3**(4) have different centers, at least one of which is not that of $Q(\beta)$.)

We have labeled in a manner invariant under both Q and L, and hence also under H. Moreover, this labeling is correct with respect to H (cf. **3.1.3**), and hence preserves the relation of "collinearity" of points.

Time: $O(\mu[qe + qd])$ to label one point, testing at most one generator a per transvection group when labeling points lying in Q.

Remark. Note that, for any $u, v \in Q$, the point $(Q(\gamma)^u)^v$ is labeled by $\langle 0, \ldots, 0, 1 \rangle ((uv)\lambda^{\bullet}) = (\langle 0, \ldots, 0, 1 \rangle u\lambda^{\bullet}) v\lambda^{\bullet}$, which is the image under $v\lambda^{\bullet}$ of the label of $Q(\gamma)^u$.

3.5.2   The homomorphism λ

We now construct a homomorphism λ from G onto $\mathrm{PSL}(V)$.

We have a basis $\mathcal{B} = \{b_1, \ldots, b_{d-1}\}$ of the $\mathbb{F}$–space Q; each member is in $\mathcal{S}^*$, and $b_i = b_1^{c^{i-1}}$. Recall that the only points we can deal with are those given explicitly as G–conjugates of known points. Therefore, we first find $l \in L$ such that $t^l = b_1$, so that $b_i \in Q(\beta_i) := Q(\alpha)^{lc^{i-1}}$ for each i. Namely, write t as a linear combination $\sum k_i b_i$ using **3.4.3**(4); let $l' \in L\lambda_L$ be any transvection such that $e_1^{l'} = \sum k_i e_i$; then use $l := l'^{-1}\lambda_L^{-1}$. (For, using Lemma 3.14(ii) and (3.4) we have $b_1 = r(e_1)\lambda^{\bullet-1} = \{r(\sum k_i e_i)\lambda^{\bullet-1}\}^{l'^{-1}\lambda^{\bullet-1}} = (\sum k_i b_i)^l$.)

Also, write $Q(\beta_d) := Q(\gamma)$; this is not on Q. If $\langle v_i \rangle$ is the label of $Q(\beta_i)$, then $v_1, \ldots, v_d$ is the *standard basis* of $V = \mathbb{F}^d$. Namely, this is clear for v_d in view of **3.5.1**, and for $1 \le i \le d - 1$ it follows from (3.4) and **3.5.1** since $b_i \lambda^\bullet = r(e_i)$ by Lemma 3.14(ii).

We need one further vector v_0 in $\mathbb{F}^d$ in "general position" with respect to the d vectors v_i: let $Q(\beta_0) := Q(\gamma)^{b_0}$, where $b_0 := \prod_1^{d-1} b_i$. By **3.5.1**, the label of $Q(\beta_0)$ is $\langle v_d \rangle(\prod_1^{d-1} b_i)\lambda^\bullet = \langle v_d \rangle r(\sum_1^{d-1} e_i) = \langle v_0 \rangle$, where $v_0 := \sum_1^d v_i$.

Now suppose that we are given $x \in G$. In order to obtain a matrix $\hat{x}$, use **3.5.1** to find the label $\langle w_i \rangle$ of $Q(\beta_i)^x$ for each $i \ge 0$. Then $\hat{x}$ sends $\langle v_i \rangle$ to $\langle w_i \rangle$. For $i \ge 0$ write $v_i \hat{x} = k_i w_i$ for nonzero scalars $k_i \in \mathbb{F}$, where $k_0 w_0 = v_0 \hat{x} = (\sum_1^d v_i)\hat{x} = \sum_1^d k_i w_i$ uniquely determines the k_i up to a scalar.

Now the required homomorphism λ is defined by $x \mapsto \hat{x}$ modulo scalar matrices for each $x \in G$. This scalar ambiguity is essential when $G \cong \mathrm{PSL}(d, q)$, but when $G \cong \mathrm{SL}(d, q)$ it will disappear once we have proved Proposition 3.18.

Correctness: We have assumed that there is an epimorphism $\lambda^{\#}: G \to \mathrm{PSL}(d, q)$. We will show that, after suitably modifying $\lambda^{\#}$, it coincides with λ modulo scalar matrices. We have already performed a modification in order to identify $\lambda^\bullet = \lambda^{\#}|_H$ in Lemma 3.14(i) and its proof.

If $b \in Q$ then $Q(\gamma)^b \lambda^{\#} = (Q(\gamma)\lambda^{\#})^{b\lambda^{\#}}$ fixes a unique 1–space $\langle 0, 1 \rangle (b\lambda^\bullet) = \langle b, 1 \rangle$ of $\mathbb{F}^d$ (cf. (3.15)). This is exactly our label for the point $Q(\gamma)^b$.

Now consider any point $Q(\beta)$ on Q. Let $1 \ne b \in [Q, Q(\beta)]$, so that $Q(\gamma)$ and $Q(\gamma)^b$ are labeled $\langle 0, 1 \rangle$ and $\langle b, 1 \rangle$, respectively. Then $Q(\beta)$ must be labeled $\langle b, 0 \rangle$ since the 1–spaces determined by $Q(\gamma)\lambda^{\#}$, $Q(\gamma)^b\lambda^{\#}$ and $Q(\beta)\lambda^{\#}$ are collinear.

Consequently, we have shown that, up to a semilinear transformation of V, *the labeling of the 1–spaces of V is uniquely determined by the action of H on $\{Q^g \mid g \in G\}$*. Since G (or rather, $G\lambda^{\#}$) also must determine a labeling of these 1–spaces, it follows that our labeling is G–invariant. This means that, for any point $Q(\alpha)^y$ and any $x \in G$, the label for $Q(\alpha)^{yx}$ is obtained from that of Q^y by applying $x\lambda^{\#}$. In particular, $\langle v_i \hat{x} \rangle = \langle w_i \rangle = \langle v_i \rangle(x\lambda^{\#})$ for $0 \le i \le d$. These conditions for $1 \le i \le d$ determine $\hat{x}$ up to a diagonal matrix, and the condition for $i = 0$ shows that this diagonal matrix is a scalar matrix. Thus, the matrix $\hat{x}$ produced by our algorithm is just the matrix induced by $x\lambda^{\#}$ on $\mathbb{F}^d$ (modulo scalars), as required.

The same type of argument will appear implicitly later in this section and in Sections 4–6.

Time: $O(\mu qd + \mu d^2 \log q)$ to find l and the $Q(\beta_i)$; then $O(d \cdot \mu[qe + qd])$ to find $\hat{x} = x\lambda$. However, $\mathcal{S}^*\lambda$ can be found faster than just proceeding in this manner for all $x \in \mathcal{S}^*$, as we will see after the following observation:

Consistency of λ and λ_L. For any $x \in L$,

$$x\lambda = \begin{pmatrix} x\lambda_L & O \\ O & 1 \end{pmatrix} \tag{3.16}$$

modulo scalars. Namely, for $b \in Q$ we have $b^x = b^{\tilde{x}} = b(x\lambda_L)$ by Lemma 3.14(ii). By **3.5.1**, $(Q(\gamma)^b)^x = (Q(\gamma)^{x^{-1}})^{bx} = Q(\gamma)^{b^x}$ says that $\langle \dot{b}, 1 \rangle \hat{x} = \langle \dot{b^x}, 1 \rangle = \langle \dot{b(x\lambda_L)}, 1 \rangle$. Since $L\lambda$ must fix $\langle v_d \rangle$, this proves (3.16).

We also claim that we may assume that $b\lambda = b\lambda^\bullet$ for all $b \in Q$. Namely, b fixes every point on Q, so that $b\lambda$ acts as a scalar on the hyperplane $\langle v_1, \ldots, v_{d-1} \rangle$ of V. Moreover, by **3.5.1**, $Q(\gamma)^b$ is labeled $\langle v_d \rangle (b\lambda^\bullet)$. Thus, $\hat{b} = b\lambda$ and $b\lambda^\bullet$ agree up to a scalar, which we may assume is 1 for all $b \in Q$.

Recursively finding $\mathcal{S}^*\lambda$:   By definition, $\mathcal{S}^*\lambda$ is the union of three sets: $\mathcal{S}^*(1)\lambda = \mathcal{S}_L^*\lambda$, $\mathcal{S}^*(2)\lambda$ coming from the $O(ed)$ generators of the groups Q and $Q(\alpha)$, and $\mathcal{S}^*(3)\lambda$ coming from the $O(e)$ generators of J. By (3.16), $\mathcal{S}_L^*\lambda$ can be obtained from $\mathcal{S}_L^*\lambda_L$ by simple bordering in $O(|\mathcal{S}_L^*|d)$ time. $\mathcal{S}^*(3)\lambda$ can be found using $O(ed)$ applications of **3.5.1**. Since $b_i^{s_i^j}\lambda = r(\rho^j e_i)$ by (3.15) and **3.4.2**, we obtain $Q \cap \mathcal{S}^*(2)$ in $d^2|\mathcal{S}^*(2)|$ time, and similarly for $Q(\alpha) \cap \mathcal{S}^*(2)$; thus $\mathcal{S}^*(2)$ is obtained in time d^3, which is $O(\mu d)$.

Time: $O(ed\mu[qe + qd])$ to find $\mathcal{S}^*\lambda$ recursively, in addition to the time to find $\mathcal{S}_L^*$.

Finding $\mathcal{S}^*\lambda$ in Theorem 1.1′:   We now have found $x\lambda$ up to scalars for each $x \in \mathcal{S}^*$. This is all we need for Theorem 1.1. However, for Theorem 1.1′ we need to eliminate the ambiguity "modulo scalars". For this, recall that $\mathcal{S}^*$ consists of elements of order p, so simply replace each matrix $x\lambda$ by $(x\lambda)^{1-q}$ for each $x \in \mathcal{S}^*$. This has the effect of killing any nontrivial scalars while not changing the coset $(x\lambda)Z(\mathrm{SL}(d, q))$. We will use powers in this manner several more times, later in this section and in Sections 4–6. **Time:** $O(d^3 \cdot d^2 e \log q)$.

In the context of Theorem 1.1′ we will find the precise image $x\lambda$ of *any* $x \in G$ at the end of **3.5.3**: the image will no longer be only modulo scalars. For this we will need straight-line programs from $\mathcal{S}^*$ to elements of G (cf. **3.5.3**).

The element m: For use in the next section we append an additional element m to $\mathcal{S}^*$, where

$$m^{-1} := l\lambda = \begin{pmatrix} l\lambda_L & O \\ O & 1 \end{pmatrix} = \begin{pmatrix} l' & O \\ O & 1 \end{pmatrix},$$

where $l = l'\lambda_L^{-1} \in L$ was defined above. We also compute the GF(p)–basis $(\mathcal{B}'_p\lambda)^{m^{-1}} = (\mathcal{B}'_p\lambda)^{m^{-1}}$ of $Q(\beta_1)\lambda$. **Time:** $O(d^3e)$.

3.5.3   Theorems 1.1(iii,iv) and 1.1'(iii,iv)

Proposition 3.17 *In deterministic $O(d^4 \log q)$ time, given $h \in \mathrm{GL}(d, q)$,*

(a) *it can be determined whether or not $h \in G\lambda$ (modulo scalars), and*

(b) *if $h \in G\lambda$ (modulo scalars) then a straight-line program of length $O(d^2 \log q)$ can be found from $\mathcal{S}^*\lambda$ to h (only modulo scalars if $G \cong \mathrm{PSL}(d, q)$), after which $h\lambda^{-1}$ can be found in additional $O(\mu d^2 \log q)$ time.*

Moreover,

(c) *in Theorem 1.1', $\mathrm{Z}(G)$ can be found in $O(\mu d^2 \log q)$ time.*

Proof. The membership test for (a) just uses a determinant calculation. Except for the last part of (b) we will only consider $G\lambda$, not G. We know all elements of $\mathcal{S}^*\lambda = \mathcal{S}^*(1)\lambda \cup \mathcal{S}^*(2)\lambda \cup \mathcal{S}^*(3)\lambda$. The groups $\langle \mathcal{S}^*(1)\lambda \rangle$ and $\langle \mathcal{S}^*(3)\lambda \rangle$ behave exactly as $\langle \mathcal{S}^*(1) \rangle$ and $\langle \mathcal{S}^*(3) \rangle$ did; we will use properties of these groups, but we will have much faster algorithms available than we had before: most of the time expended in **3.3.1–3.4.4** was concerned with relating L and $L\lambda_L$, whereas in the present situation we only need to relate $L\lambda_L$ to itself using the identity map. Therefore, as indicated below, we can perform Lemmas 3.11 and 3.13 faster.

We will prove the assertions in (b) concerning Theorems 1.1 and 1.1' simultaneously, writing elements of $G\lambda$ as matrices in either case. We proceed recursively, the base cases $d = 2$ and 3 appearing in **3.6.1** and **3.6.3**. Hence we will assume that $d \geq 4$, but in fact everything we describe works for $d = 3$ as well. We begin by recalling the various ingredients at our disposal.

(i) $v_1, \ldots, v_d$ is the standard basis of $V = \mathbb{F}^d$ in **3.5.2**, where $\langle v_1 \rangle$ is the label of $Q(\beta_1)$ and $\langle v_d \rangle$ is the label of $Q(\gamma)$. All matrices are written using this basis.

(ii) $Q\lambda$ is 1 on the hyperplane $W = \langle v_1, \ldots, v_{d-1}\rangle$, and any element of $Q\lambda$ looks like $y = \begin{pmatrix} I & O \\ k_1 \cdots k_{d-1} & 1 \end{pmatrix} = \prod_1^{d-1} k_i b_i \lambda$.

In $O(de)$ time we can look up these coefficients k_i in terms of the basis $\{M_\zeta\}$ of $\mathbb{F}$ using the tables stored in **2.3**, and then we can write a straight-line program of length $O(d\log q)$ from $\mathcal{B}_p\lambda \subset \mathcal{S}^*(2)\lambda$ to y in $O(d\log q)$ time.

(iii) In **3.5.2** we saw that $H\lambda$ consists of all $\begin{pmatrix} B & O \\ * & 1 \end{pmatrix}$ with $\det B = 1$ (cf. (3.16)).

(iv) The matrix $m = (l\lambda)^{-1} \in \mathcal{S}^*\lambda$, computed at the end of **3.5.2**, conjugates $Q(\beta_1)\lambda$ to $Q(\alpha)\lambda$. As in (ii), since $Q(\beta_1)\lambda$ is 1 on $V/\langle v_1\rangle$, any given element $y \in Q(\beta_1)\lambda$ can be expressed as an $\mathbb{F}$-linear combination of $(\mathcal{B}'\lambda)^{m^{-1}}$ in $O(d^3)$ time. In $O(d\log q)$ further time a straight-line program of length $O(d\log q)$ can be obtained from $(\mathcal{B}'_p\lambda)^{m^{-1}}$ to y.

(v) The element $j(\gamma) \in \mathcal{S}^*(3)$ in **3.2.2**(iv) conjugates $Q(\alpha)$ to $Q(\gamma)$, so that $k := j(\gamma)\lambda \in \mathcal{S}^*(3)\lambda$ conjugates $Q(\alpha)\lambda$ to $Q(\gamma)\lambda$.

(vi) We have a certain element $j \in J$ in **3.2.2**(v) normalizing T and fixing γ, and a straight-line program of length $O(\log q)$ from $\mathcal{S}^*(3)$ to j and hence another one from $\mathcal{S}^*(3)\lambda$ to $j\lambda$.

We have $[V, Q(\beta_1)\lambda] = \langle v_1\rangle$ and $[V, Q(\gamma)\lambda] = \langle v_d\rangle$. In general, we can identify any "point" $Q(\delta)\lambda$ with the point $[V, f]$ of V whenever $1 \neq f \in Q(\delta)\lambda$. Clearly a 1–space δ is on W if and only if the last coordinate of one nonzero vector in δ is 0. If $g_1, g \in G\lambda$ and $Q(\delta_1)\lambda = [Q(\beta_1)\lambda]^{g_1}$ are given, then $[Q(\beta_1)\lambda]^{g_1 g}$ can be found as δ_1^g, using matrix multiplication in $O(d^2)$ time. These are the only types of points we will encounter.

We will also need to deal with hyperplanes of V. If a hyperplane "is" a conjugate $[Q\lambda]^g$ of $Q\lambda$, $g \in G\lambda$, then its equation is $(x_i)\{v_d g^{-t}\}^t = 0$ in coordinate form, and hence can be found in $O(d^3)$ time (where g^{-t} is the inverse transpose of the matrix g). This hyperplane is on $Q(\beta_1)\lambda$ if and only if the first coordinate of $\{v_d g^{-t}\}^t$ is 0.

With all of these preliminaries in mind, we can now begin our algorithm. Let $h \in G\lambda$. By multiplying h with an appropriate scalar we may assume that $\det(h) = 1$ and hence that $h \in G\lambda$.

We first reduce to the case in which h leaves W invariant. We already have a point $Q(\gamma)\lambda = [Q(\alpha)\lambda]^k = [Q(\beta_1)\lambda]^{mk}$ not on $Q\lambda$. Find $u \in \mathcal{B}\lambda$ such that $[Q(\beta_1)\lambda]^{mku}$ is not on $[Q\lambda]^h$. If the hyperplanes $[Q\lambda]^{(mk)^{-1}}$ and $[Q\lambda]^{(mkuh^{-1})^{-1}}$, which are not on $Q(\beta_1)\lambda$, have respective equations $(x_i)(1, w_1)^t = 0$ and $(x_i)(1, w_2)^t = 0$, then $y := \begin{pmatrix} 1 & O \\ w_2^t - w_1^t & I \end{pmatrix}$ is the unique element in $Q(\beta_1)\lambda$ sending $[Q\lambda]^{(mk)^{-1}}$ to $[Q\lambda]^{(mkuh^{-1})^{-1}}$. By (iv), y can be obtained from our GF(p)–basis $(\mathcal{B}_p'\lambda)^{m^{-1}}$ for $Q(\beta_1)\lambda$ by a straight-line program of length $O(d \log q)$. This produces a straight-line program to y^m from $\mathcal{B}_p' \subseteq \mathcal{S}^*(2)\lambda$. We have $k \in \mathcal{S}^*(3)\lambda$ and $u \in \mathcal{S}^*(2)\lambda$. Since $(mk)^{-1}y(mkuh^{-1}) = k^{-1}y^m kuh^{-1}$ normalizes $Q\lambda$, if we can find a straight-line program of length $O(d^2 \log q)$ from $\mathcal{S}^*\lambda$ to this element then we can also obtain one of length $O(d^2 \log q)$ from $\mathcal{S}^*\lambda$ to h.

Thus, we may now assume that h normalizes $Q\lambda$ and hence fixes W. Write $h = \begin{pmatrix} \bar{h} & O \\ * & a \end{pmatrix}$, where $\bar{h}$ is a $d-1 \times d-1$ matrix and $a \in \mathbb{F}^*$. By **3.2.2**(v) together with (iii), $j\lambda \in J\lambda$ has the form $j\lambda = \begin{pmatrix} \bar{j} & O \\ O & a' \end{pmatrix}$, where $\bar{j}$ is a $d - 1 \times d - 1$ matrix, $a' \in \mathbb{F}^*$ has order $q - 1$, and $\det(j\lambda) = 1$; and by (vi) we have a straight-line program of length $O(\log q)$ from our generators to $j\lambda$. Use the exponents in the Zech table in **2.3** to find a power j' of $j\lambda$ such that hj' has the form $\begin{pmatrix} B & O \\ * & 1 \end{pmatrix}$ with $\det B = 1$. Recursively find a straight-line program from $\mathcal{S}_L^*\lambda$ to $b := \begin{pmatrix} B^{-1} & O \\ O & 1 \end{pmatrix}$. Then $hj'b = \begin{pmatrix} I & O \\ * & 1 \end{pmatrix}$ is in $Q\lambda$, and by (ii) can be obtained from $\mathcal{S}^*(2)\lambda$ by a straight-line program of length $O(d \log q)$. Hence, in view of our recursive call, we can obtain h from $\mathcal{S}^*\lambda$ using a straight-line program of length $O(d^2 \log q)$ (whether $G\lambda$ is PSL(d, q) or SL(d, q)).

Time: The element $u \in \mathcal{B}\lambda$ is found in $O(d \cdot d^2)$ time, while y and y^m, and straight-line programs to them, can be found in $O(d^3 + d \log q)$ time. Also j' can be computed in $O(d^3 \log q)$ time using $O(\log q)$ matrix multiplications. In view of the recursive call, a straight-line program from $\mathcal{S}^*\lambda$ to h is obtained in $O(d \cdot d^3 \log q)$ time. Moreover, the length of such a program is $O(d \cdot d \log q)$. This proves the first part of (b).

Finally, in order to find $h\lambda^{-1}$ in (b), in G mirror the straight-line program for h using the elements of $\mathcal{S}^*$ in place of those of $\mathcal{S}^*\lambda$. The resulting straight-line program ends with $h\lambda^{-1}$. Then in G perform $O(d^2 \log q)$ multiplications in $O(\mu d^2 \log q)$ time.

(c) We have $\mathrm{Z}(G\lambda) = \langle\mathrm{diag}(\varepsilon, \varepsilon, \ldots, \varepsilon)\rangle$, where $\varepsilon := \rho^{(q-1)/(d,q-1)}$ in the notation of **2.3**. Hence, apply (b) to find a straight-line program to $\mathrm{Z}(G\lambda)$ and then also one to $\mathrm{Z}(G)$. $\square$

Remark. The algorithm could be sped up by a more standard use of Gaussian elimination. However, we have tried to set the pattern to be used in later sections. The next argument, which uses this pattern, is more complicated than the corresponding ones later, when only one conjugacy class of subgroups of G arises (i.e., when the stabilizer of a point is also the stabilizer of a hyperplane).

Proposition 3.18 *In deterministic $O(\mu[qe + qd + d^2\log q])$ time, a straight-line program of length $O(d^2\log q)$ can be found from $\mathcal{S}^*$ to any given $g \in G$. (Hence, the corresponding straight-line program can be written from $\mathcal{S}^*\lambda$ to $g\lambda$, and then $g\lambda$ can be computed in additional $O(d^3 \cdot d^2\log q)$ time.)*

Proof. Again we begin by reducing to the case in which g normalizes Q. Suppose that $Q^g \neq Q$ (using the test in **3.3.2**). Recall that $k := j(\gamma) \in J$ is an element of $\mathcal{S}^*$ moving $Q(\alpha)$ to a point $Q(\gamma)$ not on Q (cf. **3.2.2**(iv)). Find $u \in \mathcal{B} \subset Q$ such that $Q(\alpha)^{kug^{-1}}$ is not on Q. Then $Q(\alpha)$ is on neither $Q^{k^{-1}}$ nor $Q^{(kug^{-1})^{-1}}$. Use Lemma 3.11(ii) and the test in **3.3.2** to find $y \in Q(\alpha)$ such that $(Q^{k^{-1}})^y = Q^{(kug^{-1})^{-1}}$. Then $k^{-1}ykug^{-1}$ normalizes Q. Use **3.4.3**(4) to obtain a straight-line program of length $O(d\log q)$ from the generators of $Q(\alpha)$ to y. Thus, if we can find a straight-line program of length $O(d^2\log q)$ from $\mathcal{S}^*$ to $k^{-1}ykug^{-1}$ then we can obtain one from $\mathcal{S}^*$ to g also of length $O(d^2\log q)$.

Now assume that $Q^g = Q$. Use **3.4.3**(5) to find matrices $\tilde{g}$ and $\tilde{j}$ representing elements of $\mathrm{GL}(d-1, q)$ induced by the actions of g and j on Q. By Lemma 3.5, $\det \tilde{g}$ is a power of $\det \tilde{j}$. Use the exponents in our Zech table in **2.3** to find $j' \in \langle j \rangle$ such that $\det(\tilde{g}\tilde{j'}) = 1$, and hence $\tilde{g}\tilde{j'}$ is a matrix in $\mathrm{SL}(d-1, q)$. Use Proposition 3.17 (applied to L) to find a straight-line program of length $O(d^2\log q)$ from $\mathcal{S}^*(1) = \mathcal{S}_L^*$ to $r := (\tilde{g}\tilde{j'})\lambda_L^{-1} \in L$. Then $\tilde{g}\tilde{j'} = r\lambda_L = \tilde{r}$ by Lemma 3.14(ii), so that $c := gj'r^{-1}$ has matrix I on Q, and hence lies in $\mathrm{C}_G(Q) = Q\mathrm{Z}(G)$. Then $c = c^{1-q}c^q$ with $c^{1-q} \in Q$ and $c^q \in \mathrm{Z}(G)$. Use **3.4.3**(4) to find a straight-line program expressing c^{1-q} in terms of our generators of Q. The algorithm ends here in the case of Theorem 1.1. For Theorem 1.1' we use Proposition 3.17(c) to obtain c^q from $\mathcal{S}^*$. Consequently, we have obtained g from $\mathcal{S}^*$, as required.

The final assertion is proved as in the preceding proposition.

Time: The dominating terms for finding a straight-line program are $O(\mu[qe + qd])$ to find y and $O(\mu d^2\log q)$ to find r. Then $g\lambda$ is obtained using $O(d^2\log q)$ matrix multiplications, each taking $O(d^3)$ time. $\square$

3.6   Small dimensions and total time

In this section we will complete the proof of Theorems 1.1 and 1.1' for the groups $\mathrm{PSL}(d,q)$ and $\mathrm{SL}(d,q)$, *assuming* that we know d and q (except for the additional verification using a presentation; cf. **7.2.2**). We first need to consider the groups $\mathrm{PSL}(2,q)$, $\mathrm{SL}(2,q) \circ \mathrm{SL}(2,q)$ (the central product) and $\mathrm{PSL}(3,q)$. These can all be handled reasonably quickly by brute force (cf. **3.2.3**).

3.6.1 $\mathrm{PSL}(2,q)$

The case $G \cong \mathrm{PSL}(2,q)$ or $\mathrm{SL}(2,q)$ was not yet considered in this section. *Here we handle it in $O(\xi qe + \mu q \log^2 q)$ time.* We may assume that $q > 9$. Note that, unlike every other situation in this paper, here we will have no smaller dimensional subgroup L to rely on.

Choose up to $6q$ elements of G in order to find, with probability $> 1 - .01 - .01$, an element t of order p and an element s such that one of the following holds: $|s| = q-1$; $|s| = (q-1)/2 \equiv 1$ (mod 2); or both $|s| = (q-1)/2 \equiv 0$ (mod 2) and $[s^{(q-1)/4}, t] \neq 1$. (Note that $|t^G| = (q^2-1)/(2,q-1)$. An element of G has order p with probability $\geq 1/q$, and has order $(q-1)/(2,q-1)$ modulo $\mathrm{Z}(G)$ with probability $\geq 1/(q-1)$.)

Now choose up to $128(q+1)e$ conjugates of t; a conjugate commutes with t with probability $(q-1)/(q^2-1)$. An application of Lemma 2.8 of type **2.5**(1) with the parameters $r = 1/(q+1), t = 128(q+1)e, \varepsilon = 1/4$, shows that we *get at least $96e$ transvections which commute with t*, and hence lie in the same transvection group T, with probability $> 1 - 1/50$. We claim that *the $96e$ transvections $u_1, \ldots, u_{96e}$ generate T with probability $> 1 - 1/50$.* For, if $\langle u_1, \ldots, u_{i-1}\rangle \neq T$ then $\mathrm{Prob}(u_i \notin \langle u_1, \ldots, u_{i-1}\rangle) \geq ((p^e-1)/2 - p^{e-1})/(p^e-1) \geq 1/8$ for p odd and $\mathrm{Prob}(u_i \notin \langle u_1, \ldots, u_{i-1}\rangle) \geq ((2^e-1) - 2^{e-1})/(2^e-1) > 1/8$ for $p = 2$. Hence an application of Lemma 2.8 of type **2.5**(2) with the parameters $r = 1/8, t = 96e, \varepsilon = 7/8$, shows that the u_i's generate T with the stated probability. The order of the group generated is tested, and a basis $\{t_1, \ldots, t_e\}$ is obtained, using Lemma 2.1. Thus, we have obtained T with probability $> 1 - 0.02 - 2/50 > 1 - 1/10$ (compare [CLG2]).

For any given $g \in G$ note that we can find $T^g = \langle t_1^g, \ldots, t_e^g\rangle$ in $O(\mu e)$ time.

Find a generator j of G such that $[t^j, t] \neq 1$, and hence j does not normalize T. Although G acts 2–transitively on the set $\Omega := \{T\} \cup \{T^{ju} \mid u \in T\}$ of size $q+1$ in a very familiar manner, we do not know how to determine the permutation action of the generators of G in a direct fashion. Instead we will construct the 2–dimensional (projective) matrix representation for G using the fact that each member of Ω contains exactly one member of $Y := \{t, t^{ju} \mid u \in T\}$.

First we need to construct a field GF(q). For each $y \in Y$, test whether $[y^s, y] = 1$ in order to find the two members T^a, T^b of Ω normalized by s. By conjugating s by some element of G that sends $\{T^a, T^b\}$ to $\{T, T^j\}$, we may assume that s normalizes T and T^j. (Namely, if $T^a \neq T$, find $u \in T$ such that $T^{au} = T^j$; then find $v \in T$ with $T^{buj^{-1}v} = T^j$; and finally replace s by $s^{uj^{-1}v}$.) Then s induces an automorphism $\tilde{s}$ of order $q - 1$ or $(q - 1)/2$ on T. The desired field GF(q) is the GF(p)–span of $\tilde{s}$. Obtain the set GF$(q)^*$ by using Lemma 2.4.

We use Lemma 2.3 to identify GF(q) and $\mathbb{F}$, letting ρ be as in **2.3**. Applying $\rho^k \in \mathbb{F}$ to some $u \in T$ takes $O(\mu \log q)$ time, by writing ρ^k as a GF(p)-linear combination of $1, \rho, \ldots, \rho^{e-1}$ (recall that the coefficients in this linear combination were stored in **2.3**).

However, if this is the first time during the recursion in Section 3 that PSL$(2, q)$ or SL$(2, q)$ is processed, so that $\mathbb{F}$ has not yet been constructed (cf. **2.3**), then we have to search in GF$(q)^*$ for an element of order $q - 1$ needed in **2.3**. If q is even then we just use conjugation by s. If q is odd then let $q - 1 = 2^k m$, with m odd. We have an endomorphism ε_1 of order $(q - 1)/2$ at hand (conjugation by s), and another endomorphism ε_2 of order $2^k m'$ with m' odd, constructed by Lemma 2.4. Then $\varepsilon := \varepsilon_1^{2^{k-1}} \varepsilon_2^m$ has order $q - 1$.

Once an element ε of order $q - 1$ is found, we can write an $e \times e$ matrix with entries from GF(p) for ε in the basis $t, t\varepsilon, \ldots, t\varepsilon^{e-1}$ by expressing $t\varepsilon^e$ as a linear combination of the basis vectors. $t\varepsilon^e$ can be computed in $O(\mu e)$ time if q is even and $O(\mu q e)$ time if q is odd and then $t\varepsilon^i$, $0 \leq i \leq e$, can be found in the listing of T in $O(\mu q e)$. Since, during listing of T, we stored coefficients for each element of T in a fixed basis, the linear combination for $t\varepsilon^e$ can be computed in $O(e^3)$ time. After the matrix for ε is constructed, we proceed with the construction of the Zech table as described in **2.3**.

Using $\rho \in \mathbb{F} = $ GF(q) in **2.3**, we re-list $T - \{1\}$ as $\{t\rho^i \mid 0 \leq i < q-1\}$ and $Y - \{t\}$ as $\{t^{j(t\rho^i)} \mid 0 \leq i < q-1\}$.

Define $\mathcal{S}^* := \langle t\rho^i, (t\rho^i)^j \mid 0 \leq i < e \rangle$, so that $G = \langle \mathcal{S}^* \rangle$ and $\mathcal{S}^*$ consists of transvections.

Using coordinates within $\mathbb{F}^2$, label $T = T_\infty$ as $\langle 1, 0 \rangle$, $T^j = T_0$ as $\langle 0, 1 \rangle$, $T^{jt} = T_1$ as $\langle 1, 1 \rangle$, and $T^{j(t\rho^i)}$ as $\langle \rho^i, 1 \rangle$ whenever $1 \leq i < q - 1$. Write $v_\infty = (1, 0), v_0 = (0, 1), v_1 = (1, 1) \in \mathbb{F}^2$.

For any given $x \in G$ we obtain $\hat{x} \in \mathrm{GL}(2, q)$ as follows. Determine the member of Ω equal to T_∞^x as the one containing the element of Y commuting with t^x. Find T_0^x and T_1^x in the same way. Now find the label $\langle w_i \rangle$ of T_i^x, so that $\hat{x}$ sends $\langle v_i \rangle$ to $\langle w_i \rangle$ for $i \in \{\infty, 0, 1\}$. As in **3.5.2** this uniquely determines the matrix $x\lambda := \hat{x}$ up to scalars. (N.B.—Unlike the situation in the rest of this section, there is no consistency required with lower-dimensional linear groups. Note, however, that λ was uniquely determined only due to our labeling of the field. If λ is followed by an automorphism of $\mathrm{PSL}(2, q)$, such as a field automorphism, then another homomorphism is obtained, even though the resulting group of matrices is the same.)

Correctness: We have assumed that there is an epimorphism $\lambda^\#: G \to \mathrm{PSL}(2, q)$; as usual we will write its images as matrices modulo scalars (actually modulo $\pm I$). We will show that $\lambda^\#$ can be modified so as to coincide with our λ modulo scalars.

By following $\lambda^\#$ by an automorphism of $\mathrm{PGL}(2, q)$ we may assume that $T\lambda^\#$ consists of all $\begin{pmatrix} 1 & 0 \\ * & 1 \end{pmatrix}$, $T_0\lambda^\#$ consists of all $\begin{pmatrix} 1 & * \\ 0 & 1 \end{pmatrix}$, and $t\lambda^\# = \begin{pmatrix} 1 & 0 \\ 1 & 1 \end{pmatrix}$. Then $T\lambda^\#$ consists of all $(t\lambda^\#)\alpha = \begin{pmatrix} 1 & 0 \\ \alpha & 1 \end{pmatrix}$, $\alpha \in \mathbb{F}$, where $[\mathbb{F}^2, (T_0\lambda^\#)^{(t\lambda^\#)\alpha}] = \langle \alpha, 1 \rangle$ is just our label for $T_0^{t\alpha}$. Also $[\mathbb{F}^2, T\lambda^\#] = \langle 1, 0 \rangle$ is our label for $T\lambda^\#$.

If $x \in G$ then $x\lambda^\#$ acts on $\Omega\lambda^\# = \{T\lambda^\#\} \cup (T_0\lambda^\#)^{T\lambda^\#}$ exactly as it does on the points of $\mathbb{F}^2$, and hence produces exactly the same matrix $\hat{x}$ as above (modulo scalars). Thus, λ and $\lambda^\#$ coincide modulo scalars, as required.

Reliability: $> 1 - 1/2 \cdot 2^2$.

Time: $O(\xi qe + \mu q \log^2 q)$. It takes $O(q[\xi + \mu \log^2 q])$ time to find t and s (note the last paragraph of **2.2.3**), and $O(\xi qe + \mu qe)$ to find and list T using Lemma 2.1. It also took $O(\mu q \log^2 q)$ time to find the generator ε. Finding j takes $O(\mu |\mathcal{S}|)$ time, which we are assuming is $O(\xi)$ (cf. **2.2.2**). In Lemmas 2.3 and 2.4 we have $\nu = O(\mu)$. Note that, *for any given $x \in G$, $x\lambda$ has been computed in $O(\mu q)$ time modulo scalars.*

Images of elements of order p. If $|x| = p$, replace $x\lambda$ by $(x\lambda)^{1-q}$ in order to obtain a matrix of order p: nontrivial scalars have been killed. In particular, we now have $\mathcal{S}^*\lambda$ and $T\lambda$. Moreover, for any $x \in G$ for which we have a straight-line program from $\mathcal{S}^*$ (see below), we can now obtain an unambiguous image $x\lambda$. Similarly, if $y \in \mathrm{SL}(2, q)$ and we have a straight-line program from $\mathcal{S}^*\lambda$ to y (see below), we can obtain a preimage $y\lambda^{-1}$.

Normalizers of $T\lambda$ and T. We need to find preimages of elements of

PSL$(2,q)$ or SL$(2,q)$, as well as the straight-line programs in Theorems 1.1 and 1.1′. However, first we construct additional ingredients: the normalizers of $T\lambda$ and T. Note that, in the basis v_∞, v_0, the groups $T\lambda$ and $T^j\lambda$ consist of all $\begin{pmatrix} 1 & 0 \\ * & 1 \end{pmatrix}$ and $\begin{pmatrix} 1 & * \\ 0 & 1 \end{pmatrix}$, respectively; normalizing one of these groups is the same as fixing the 1–space $\langle v_\infty \rangle$ or $\langle v_0 \rangle$, respectively.

Let $j' := \begin{pmatrix} 0 & -1 \\ 1 & 0 \end{pmatrix} = \begin{pmatrix} 1 & 0 \\ 1 & 1 \end{pmatrix}\begin{pmatrix} 1 & 1 \\ 0 & 1 \end{pmatrix}^{-1}\begin{pmatrix} 1 & 0 \\ 1 & 1 \end{pmatrix}$; we can obtain a straight-line program of length 3 from $\mathcal{S}^*\lambda$ to j' and then also one from $\mathcal{S}^*$ to $j'\lambda^{-1}$ in order to find the latter element. Note that $T^j = T^{j'\lambda^{-1}}$.

By looking up the GF(p)-coefficients of $-\rho^{-1}$ stored with $\mathbb{F}$ in **2.3**, find a straight-line program of length $O(\log q)$ from $\mathcal{S}^*\lambda$ to

$$l' := j'\begin{pmatrix} 1 & 0 \\ -\rho^{-1} & 1 \end{pmatrix}\begin{pmatrix} 1 & \rho \\ 0 & 1 \end{pmatrix}\begin{pmatrix} 1 & 0 \\ -\rho^{-1} & 1 \end{pmatrix} = \begin{pmatrix} \rho^{-1} & 0 \\ 0 & \rho \end{pmatrix}.$$

Then $N_{G\lambda}(T\lambda) = T\lambda \rtimes \langle l' \rangle$ and $N_{G\lambda}(T\lambda) \cap N_{G\lambda}(T^j\lambda) = \langle l' \rangle$. Similarly, we obtain a straight-line program from $\mathcal{S}^*\lambda$ to any given element of $\langle l' \rangle$; and then also a straight-line program from $\mathcal{S}^*$ to $l := l'\lambda^{-1}$, where $N_G(T) \cap N_G(T^j) = \langle l \rangle$. (N.B.—$\langle l \rangle$ can contain $\langle s \rangle$ properly if $Z(G) \neq 1$. We could have searched for an element s of G of order $q-1$ in this case.)

Time: $O(\log q)$ for j' and l' and then $O(\mu \log q)$ for l.

Theorems 1.1(iv) and 1.1′(iv) (i.e., Proposition 3.17): Use a determinant to test whether the given element $h \in$ GL$(2,q)$ lies in SL$(2,q)$. Suppose that $h \in$ SL$(2,q)$. We will construct a straight-line program of length $O(\log q)$ from $\mathcal{S}^*\lambda$ to h modulo scalars. If $(T\lambda)^h \neq T\lambda$ find $u \in T\lambda$ such that $(T\lambda)^{hu} = (T\lambda)^{j'}$; write a short straight-line program from $\mathcal{S}^*$ to u, and replace h by huj'^{-1}. Now $(T\lambda)^h = T\lambda$.

Similarly, if $(T\lambda)^{j'h} \neq (T\lambda)^{j'}$ find $u \in T\lambda$ with $(T\lambda)^{j'hu} = (T\lambda)^{j'}$. Thus, we may assume that h normalizes both $T\lambda$ and $(T\lambda)^{j'}$. Now use the Zech table in **2.3** to find h as a power of l', and hence also obtain h by using a straight-line program of length $O(\log q)$ from $\mathcal{S}^*\lambda$.

Time: $O(\mu \log q)$; u is found by solving a linear equation in $\mathbb{F}$ in one unknown in $O(1)$ time.

Theorem 1.1(iii) and 1.1′(iii) (i.e., Proposition 3.18): We are given $g \in G$ and need a straight-line program from $\mathcal{S}^*$ to g. As above first arrange for g to normalize T, then for it also to normalize T^j, then to find g as a power of l, and finally to find a straight-line program to g of length $O(\log q)$ from $\mathcal{S}^*$.

Time: $O(\mu q)$ since this time we need to test all elements of T in order to make g normalize first T and then also T^j.

$Z(G)$: This is $\langle (j'\lambda^{-1})^2 \rangle$, since the only elements of G that interchange a pair of conjugates of T have order 2 mod $Z(G)$. Similarly, we obtain $Z(G\lambda)$ in terms of a straight-line program from $\mathcal{S}^*\lambda$.

A simple variation of the algorithm presented in this section produces the following deterministic version that will be used in Lemma 6.9:

Corollary 3.19 *In deterministic $O(\mu q^2 e)$ time, given a black box group $G = \langle \mathcal{S} \rangle \cong \mathrm{SL}(2,q)$ with $|\mathcal{S}| = O(e)$, and given a subgroup of order q, one can find an isomorphism $\lambda\colon G \to \mathrm{SL}(2,q)$ behaving as in* Theorem 1.1′.

Proof. Probability entered into the algorithm earlier in this section three times: finding a transvection, finding a transvection group T, and finding an element s. We are given the first two of these, hence all we need is a deterministic procedure that produces s.

List the given subgroup T of order q using Lemma 2.1, and let $1 \neq t \in T$. Find a generator $j \in \mathcal{S}$ such that $[t^j, t] \neq 1$. Then $X := \{1\} \cup t^{jT}$ is a transversal for $\mathrm{N}_G(T)$ in G.

For each $x \in X$ and $g \in \mathcal{S}$ find $x' \in X$ such that $[t^{xgx'}, t] = 1$. Then $\mathrm{N}_G(T)$ is generated by the elements xgx' (they are "Schreier generators" for this subgroup).

Similarly, T is a transversal for $\mathrm{N}_G(T) \cap \mathrm{N}_G(T^j)$ in $\mathrm{N}_G(T)$, and we can find the cyclic group $\mathrm{N}_G(T) \cap \mathrm{N}_G(T^j)$ of order $q-1$. List it and find a generator s.

Now continue with the remainder of the algorithm earlier in this section.

Time: Dominated by the time to process triples such as x, s, x'. $\square$

3.6.2 $\mathrm{SL}(2,q) \circ \mathrm{SL}(2,q)$

In **4.2.1** we will need to handle the central product $G = \langle \mathcal{S} \rangle \cong \Omega^+(4,q) \cong \mathrm{SL}(2,q) \circ \mathrm{SL}(2,q)$ when $q > 4$ is even or $q > 9$ is odd.

Choose up to $2q$ elements τ until one is found of order pz with $z > (2, q-1)$; this occurs with probability $\geq 2/q$ for a single τ and hence $> 1 - 1/2^4$ for $2q$ choices. If $t := \tau^z$, choose up to 16 pairs t_1, t_2 of random conjugates of t. Then each $\langle t, t_1, t_2 \rangle$ lies in one of the $\mathrm{SL}(2,q)$ factors; by Lemma 3.8, for a single one of our pairs $\langle t, t_1, t_2 \rangle$ is one of these factors with probability $\geq 1/4$. Test whether it is $\mathrm{SL}(2,q)$ by using up to four repetitions of **3.6.1** together with a presentation as in **7.2.2**. With probability $> 1 - 2/2^4$, for at least one of our 16 pairs we have $\langle t, t_1, t_2 \rangle \cong \mathrm{SL}(2,q)$ together with a successful test of this fact.

Now repeat in order to (probably) find the other $\mathrm{SL}(2,q)$ factor.

Reliability: $> 1 - 1/4$.

Time: $O(\xi qe + \mu q \log^2 q)$.

Note that we have also obtained isomorphisms from each factor to $\mathrm{SL}(2,q)$. Of course, the same algorithm handles $\mathrm{PSL}(2,q) \times \mathrm{PSL}(2,q)$.

3.6.3 $\mathrm{PSL}(3,q)$

We now *handle the base case* $G \cong \mathrm{PSL}(3,q)$ *or* $\mathrm{SL}(3,q)$ *in* $O(\xi qe + \mu q \log^2 q)$ *time.* We may assume that $q > 7$.

3.2.1: Test up to 128 elements τ in order to find one of $\mathrm{ppd}^{\#}(p; 2e) \cdot \mathrm{ppd}^{\#}(p; e)$– or $\mathrm{ppd}^{\#}(p; 2e) \cdot \mathrm{ppd}^{\#}(p; e) \cdot \mathrm{ppd}^{\#}(p; e/2)$–order, respectively, depending on whether e is odd or even; in addition we require that $|\tau^{2(q+1)}| > 3$, and if q is a Mersenne or Fermat prime also that $16 \big| |\tau|$. (An element has this order with probability $> 1/16$, so we will succeed with probability $> 1 - 1/2^8$.)

Let $a := \tau^{2(q+1)}$. (Then a is not a scalar, but it will have i–dimensional eigenspaces V_i for $i = 1, 2$, on each of which it induces a scalar. If $g \in G$ is such that $V_1 \neq V_1^g \not\subseteq V_2$ and $V_1 \not\subseteq V_2^g$, then $\langle a, a^g \rangle$ preserves the decomposition $(V_2 \cap V_2^g) \oplus \langle V_1, V_1^g \rangle$. Moreover, $\langle a, a^g \rangle$ then acts irreducibly on $\langle V_1, V_1^g \rangle$ and hence induces at least $\mathrm{SL}(2,q)$ on this subspace; see the proof of Lemma 3.8(i); we have included a factor $\mathrm{ppd}^{\#}(p; e/2)$ in $|\tau|$ in order to eliminate the possibility of a normal subgroup $\mathrm{SL}(2, \sqrt{q})$ arising here.)

Choose up to 16 conjugates b of a. With probability $\{(q^2 - 1)(q^2 - q - 1) / (q^2 + q + 1)q^2\} > 1/2$, for a single b we will have $A := \langle a, b \rangle \cong \mathrm{SL}(2,q) \circ \langle a \rangle$, and hence this will hold with probability $> 1 - 1/2^8$ for at least one of our choices. Find the subgroup $L \cong \mathrm{SL}(2,q)$ of A, testing this isomorphism and finding an isomorphism $\lambda_L \colon L \to \mathrm{SL}(2,q)$ by using up to three repetitions of **3.6.1**. (There are various ways to find L. Using our previous methodology, choose up to $9q$ elements of A to find one of order divisible by p, find its power t of order p, and then use $L := \langle t, t_1, t_2 \rangle$ for random A–conjugates t_1, t_2 of t; cf. Lemma 3.8(iii). Alternatively, use a Monte Carlo algorithm [BCFLS] to find $L = A'$.)

Use λ_L^{-1} to find a transvection group T in L, and let $1 \neq t \in T$.

3.2.2: Omit.

3.3.1: Choose up to $8q$ conjugates t^f of t. With probability $> 2/2q$ for a single choice, $t_1 := [t, t^f]$ will be nontrivial and commute with t, in which case t_1 will be another transvection by **3.1.2**. Let $T_1 := [T, t^f]$ and $Q := \langle T, T_1 \rangle$ (these have respective orders q and q^2).

Let $Q(\alpha) := \langle T, T_1^{f^{-1}} \rangle$. (Up to duality we can view Q as the group of all transvections having the same axis as T, and then $Q(\alpha)$ consists of all transvections whose center is the same as that of both T and $T_1^{f^{-1}}$.)

Let $Q(\beta) := Q(\alpha)^f$. Then $H := \langle Q(\alpha), Q(\beta) \rangle$ and $G'_\alpha := \langle Q, Q^{f^{-1}} \rangle$ are isomorphic to $\mathrm{ASL}(2, q)$ by **3.1.2**(v).

3.3.2: Lemma 3.11 is unchanged, while Corollary 3.12 is omitted.

3.3.3: Redefine $L := \langle T^f, T_1^{f^{-1}} \rangle \cong \mathrm{SL}(2, q)$. (This isomorphism follows from **3.1.2**: T^f and $T_1^{f^{-1}}$ have respective centers $\alpha^f = \beta$ and $\beta^{f^{-1}} = \alpha$, so that both normalize Q although neither lies in Q.)

Apply **3.6.1** to L up to three times, obtaining a field $\mathbb{F}$, an isomorphism $\lambda_L : L \to \mathrm{SL}(2, q)$, and a generating set $\mathcal{S}_L^*$ of L with probability $> 1 - 1/8^3$. Use λ_L^{-1} to find $j \in L$ of order $q - 1$ normalizing $T_1^{f^{-1}}$ and hence also $[Q, T_1^{f^{-1}}] = T$, together with a straight-line program to j of length $O(\log q)$ from $\mathcal{S}_L^*$. Use this in place of the element j appearing in **3.2.2**(v) within the remainder of the algorithm.

We still need to find a substitute for the element $j(\gamma)$ occurring in **3.2.2**(iv), taking $Q(\alpha)$ to the unique point $Q(\gamma)$ fixed by L. Note that γ corresponds to $Q^f \cap Q^{f^{-1}}$. Test all $t' \in T$ in order to find one such that $[[Q^f, t^{f^{-1}}t'], Q^f] = 1$. Let $j'(\gamma) := f^{-1}t'$. (Namely, $T^{f^{-1}}$ does not have the same center α as $T_1^{f^{-1}}$, and $T \leq L^{f^{-1}} \leq \mathrm{N}_G(Q^{f^{-1}})$. Thus T has an element t' sending $T^{f^{-1}}$ to the transvection subgroup of $Q^{f^{-1}}$ whose center is γ, and hence for which $t^{f^{-1}}t'$ normalizes Q^f.)

Let $Q(\gamma) := Q(\alpha)^{j'(\gamma)}$. Since $\langle T, T^{j'(\gamma)} \rangle \cong \mathrm{SL}(2, q)$ we see that $T^{t^{j'(\gamma)}}$ has an element $j(\gamma)$ such that $Q(\alpha)^{j(\gamma)} = Q(\gamma)$. Use $j(\gamma)$ and $Q(\gamma)$ as before (cf. **3.2.2**(iv)).

3.4–3.5: Proceed as before.

Reliability: $\geq 1 - 3/2^8 - 2/8^3 > 1 - 1/2 \cdot 3^2$.

Time: $O(\xi qe + \mu q \log^2 q)$. Note that we spend $O(\mu q \log^2 q)$ time on **3.4–3.5**, and $O(\mu qe)$ and $O(\mu \log q)$ in Propositions 3.18 and 3.17, respectively.

Corollary 3.20 *In $O(\xi qe + \mu q \log^2 q)$ time one can test whether or not the group J in **3.2.2** satisfies $J \cong \mathrm{SL}(3, q)$, and, if so, then find an isomorphism $J \to \mathrm{SL}(3, q)$ together with the elements and subgroups in **3.2.2**(i-v).*

Proof. In **3.2.2** we already have a transvection t. Find a 3–dimensional J–module using the preceding algorithm (or report failure). We assume that an isomorphism $\lambda_J : J \to \mathrm{SL}(3, q)$ is obtained. Find $t\lambda_J$. Write 3×3 matrices in order to find **3.2.2**(i-v) within $\mathrm{SL}(3, q)$. Pull these back into J using λ_J^{-1}. The timing comes from the preceding algorithm, then finding $t\lambda_J$ in $O(\mu q)$ time, and finally applying λ_J^{-1} to $O(e)$ matrices in $O(e \cdot \mu \log q)$ time (cf. **3.2.2**).

In order to test that $J \cong \mathrm{SL}(3, q)$, use a presentation as in **7.2.2**. $\square$

3.6.4   Total time and reliability

Part (i) of Theorems 1.1 and 1.1′ will be discussed later, in Section 7; (ii) is **3.5.2**; (iii) and (iv) are in **3.5.3**; and (vi) is in **3.3.3**. For (vii), by tallying and recalling the recursive call we find that the *total time is*

$$O(d\{\xi q(e + d) \log d + \mu q d^3 \log^2 q \log d\}),$$

dominated by **3.2.1** and **3.2.2**.

If $d \geq 4$ then the probability of failure before the recursive call is at most $2/4d^2$, so the entire algorithm fails with probability $\leq \sum_4^d (1/2i^2) < 1/4$. By **3.6.1** and **3.6.3**, this bound holds for the probability of failure in the cases $d = 2$ and 3 as well.

4 Orthogonal groups: $\mathrm{P\Omega}^\varepsilon(d,q)$

We now turn to the other classical groups. We begin with orthogonal groups: the analogue for them of the group Q is abelian (cf. **4.1.3**, **5.1.3**, **6.1.3**) and hence more closely resembles the situation studied earlier in **3.1.3**. Moreover, in the orthogonal case the elements of this group Q behave relatively nicely; and there are commutator relations resembling those in **3.1.2**. On the other hand, the symplectic and unitary cases are much simpler in another respect: those groups contain transvections. Not having these presents some serious obstacles in this section, and leads to poorer timing estimates and lengthier arguments. We will also have to deal with some minor difficulties with the groups $\mathrm{P\Omega}^+(8,q)$ and $\mathrm{P\Omega}^+(10,q)$ caused by the triality outer automorphism of $\mathrm{P\Omega}^+(8,q)$. Overall, from a technical standpoint this is the most difficult section of this paper.

We will prove Theorems 1.1 and 1.1′ when G is $\mathrm{P\Omega}(V)$ or $\Omega(V)$, where $\Omega(V)$ preserves a nonsingular quadratic form φ on a d–dimensional vector space V over $\mathrm{GF}(q)$. For each subspace W, as usual $W^\perp$ is the subspace of vectors perpendicular to W. The most important subspaces are those that are either nonsingular (so that $W \cap W^\perp = 0$) or totally singular (so that $\varphi(W) = 0$). We call a subspace "nondegenerate" if it has a 1–dimensional radical on which φ does not vanish (this can only occur in characteristic 2). There are three subfamilies of orthogonal groups: $\Omega^+(2n,q)$, when V has Witt index n; $\Omega^-(2n,q)$, when V has Witt index $n-1$; and $\Omega(2n+1,q)$, where q is assumed to be odd unless stated otherwise (cf. [As2, Ta, KL]; the Witt index is the maximal dimension of a totally singular subspace). We write these groups generically as $\Omega^\varepsilon(d,q)$. If α is any singular point then $\alpha^\perp/\alpha$ is an orthogonal space of dimension $d-2$ belonging to the same family as V. When the field is understood, V_{2k}^+ denotes an orthogonal $2k$–space of Witt index k for any k; the notation V_{2k}^- is used similarly for index $k-1$.

4.1 Properties of G

4.1.1 Root elements

Whenever u and w are linearly independent and $\varphi(u) = 0 = (u,w)$ there is a subgroup

$$R(\langle u,w\rangle) = \{v \mapsto v - k(v,u)w + k(v,w)u - k^2\varphi(w)(v,u)u \mid \\ k \in \mathrm{GF}(q)\} \cong \mathrm{GF}(q)^+ \qquad (4.1)$$

that depends on the 2–space $\langle u, w \rangle$ rather than on the choice of u and w. "Long" root groups $R(\langle u, w \rangle)$, for which $\varphi(w) = 0$, are the natural analogues of transvections. We will also need to use nonsingular vectors w, in which case we will follow [Ta, pp. 147-148] and also call $R(\langle u, w \rangle)$ a "root group" and its elements "root elements"; in general these are not root groups in the sense of Lie Theory, so as in [Ta, pp. 148-150] it might instead be preferable to call them "Siegel transformations". (Within Lie Theory, $\Omega^+(2m, q)$ has only one type of root groups, which we have called "long root groups"; in $\Omega(2m + 1, q)$ with q odd, in addition to the above long root groups there are also "short root groups" $R(\langle u, w \rangle)$ when $w^\perp$ has maximal Witt index; and in $\Omega^-(2m, q)$, in addition to the above long root groups there are "short root groups" of order q^2, but at least their elements are all "root elements" in the above sense.)

We note the following useful consequences of (4.1):

Lemma 4.2 *Let* $1 \neq r \in R(\langle u, w \rangle)$.

   (i) $[V, r] = \langle u, w \rangle$, $C_V(r) = [V, r]^\perp$ *contains all fixed 1–spaces of* r, *and* $R(\langle u, w \rangle)$ *induces* 1 *on* $V/[V, r]$.

   (ii) *If* $\varphi(w) \neq 0$ *then* $[V, r]$ *has a unique singular point* $[V, r]^\circ = \langle u \rangle$.

   (iii) *If* q *is odd then the totally singular 2–spaces fixed by* r *are those in* $[V, r]^\perp$ *and those lying between* z *and* $z^\perp$ *for some singular point* z *of* $[V, r]$.

4.1.2 Commutator relations

Consider long root groups R_i for $i = 1, 2$, and let $L_i = [V, R_i]$. Then R_1 and R_2 interact in one of the following ways:

   (i) $R_1 = R_2$.

   (ii) L_1 and L_2 are perpendicular, and $[R_1, R_2] = 1$.

   (iii) $\dim L_1 \cap L_2 = 1$, and $[R_1, R_2] = 1$.

   (iv) $L_1 \cap L_2 = 0$, $L_1^\perp \cap L_2 \neq 0$ and $L_2^\perp \cap L_1 \neq 0$, in which case both $[r_1, R_2]$ and $[R_1, r_2]$ are the long root group $R(\langle L_1^\perp \cap L_2, L_2^\perp \cap L_1 \rangle)$ whenever $1 \neq r_1 \in R_1$ and $1 \neq r_2 \in R_2$. Moreover, $[R_1, R_2]$ *commutes with both* R_1 *and* R_2.

   (v) $\langle L_1, L_2 \rangle$ is a nonsingular 4–space, and $\langle R_1, R_2 \rangle \cong \mathrm{SL}(2, q)$. Here, $\langle R_1, R_2 \rangle$ preserves the decomposition $V = \langle L_1, L_2 \rangle \perp \langle L_1, L_2 \rangle^\perp$, inducing $\mathrm{SL}(2, q)$ on the first summand and 1 on the second one.

4.1.3 Q

For any singular point α, let $Q = Q(\alpha)$ denote the group of all isometries inducing 1 on both α and $\alpha^\perp/\alpha$. If $\alpha = \langle u \rangle$ then Q consists of the maps

$$v \mapsto v + (v, z)u - (v, u)z \text{ with } z \in \alpha^\perp;$$

these are 1 and all of the root elements r for which $\alpha \in [V, r]$. Alternatively, if u, w is a hyperbolic pair of vectors (i.e., $\varphi(u) = \varphi(w) = 0$, $(u, w) = 1$) then Q consists of all isometries $r(z)$, $z \in \langle u, w \rangle^\perp$, defined by

$$r(z) \colon u \mapsto u, \quad v \mapsto v - (v, z)u \ \forall v \in \langle u, w \rangle^\perp, \quad w \mapsto w + z - \varphi(z)u. \quad (4.3)$$

Note that Q consists of *sparse matrices* with respect to a basis consisting of u, w and a basis of $\langle u, w \rangle^\perp$. Moreover, *$Q$ is regular on the set of all singular points not in $\alpha^\perp$*, and

$$G_\alpha = Q \rtimes G_{\alpha\gamma} \text{ where } \gamma = \langle w \rangle. \qquad (4.4)$$

We will focus on the geometry induced by G within Q. If $\zeta \in \mathrm{GF}(q)^*$ has order $(q-1)/(2, q-1)$, then the linear transformation defined by

$$s \colon u \mapsto \zeta u, \quad w \mapsto \zeta^{-1} w, \quad v \mapsto v \ \forall v \in \langle u, w \rangle^\perp \qquad (4.5)$$

lies in $\Omega(V)$ (in view of the order restriction) and $r(z)^s = r(\zeta z)$ for all z. This produces an automorphism of Q whose $\mathrm{GF}(p)$–linear span is isomorphic to the field $\mathrm{GF}(q)$, which then turns Q into a $\mathrm{GF}(q)$–space. Moreover, if $d \geq 6$ then $(G_{\alpha\gamma})'$ is the orthogonal group induced by $(G_\alpha)'$ on $\alpha^\perp/\alpha$, and the $(G_{\alpha\gamma})'$–modules $\alpha^\perp/\alpha$, $\langle \alpha, \gamma \rangle^\perp$ and Q are isomorphic. In particular, $(G_\alpha)'$ acts irreducibly on Q in view of (4.4). The quadratic form inherited on the $\mathrm{GF}(q)$–space Q is given by

$$\varphi_Q(r(z)) = \varphi(z),$$

so that the *long root elements of Q are precisely the singular "vectors" in Q*. Hence, there is, indeed, "geometry" within Q.

Lemma 4.6 *Let $g \in G$ send $u \mapsto ju$, $w \mapsto kw$ (for $j, k \in \mathbb{F}$); let g' denote the restriction of g to $\langle u, w \rangle^\perp$. If g induces by conjugation the linear transformation $\tilde{g}$ on Q, then*

(a) $r(z)^{\tilde{g}} = r(z)^g = r(jz^{g'})$ *for all $z \in \langle u, w \rangle^\perp$; and*

(b) *If $\tilde{g}$ is an isometry of Q, then*

    (i) $j = k = \pm 1$, *and*

    (ii) $\tilde{g} \notin \Omega(Q)$ *if and only if q is odd, $g = -1|_{\langle u,w \rangle} \perp g'$, and either d is odd or d is even and $-1 \notin \Omega(V)$ (in the latter event $q \equiv 3$ (mod 4) and $d \equiv 2$ (mod 4)).*

Proof. (a) We have $jk = 1$ and $w^{g^{-1}r(z)g} = w + jz^g - \varphi(z)j^2 u$ for any $r(z) \in Q$.

(b) For (i), since $\tilde{g}$ and g' are isometries, by (a) we have $\varphi_Q(v) = \varphi_Q(v^{\tilde{g}}) = j^2\varphi_Q(v)$ for all $v \in Q$, so that $j = k = \pm 1$.

For (ii), note that $1 = \det g = \det(jI_2)\det g' = \det g'$ and hence $\det \tilde{g} = \det(jI_{d-2})\det g' = j^{d-2}$. If d is even then this determinant is 1; if d is odd then $\tilde{g} \notin \Omega(Q)$ if and only if $\det \tilde{g} = -1$, which occurs if and only if $j = k = -1$.

If d is even and N denotes the spinor norm, then $1 = N(g) = N(jI_2)N(g')$ and $N(\tilde{g}) = N(jI_{d-2})N(g') = N(jI_{d-2})N(jI_2) = N(jI_d)$. Then $\tilde{g} \notin \Omega(Q)$ if and only if $j = -1$ and $N(-I_d) \neq 1$, as asserted in (ii). $\square$

Further properties of Q.

Lemma 4.7 *Suppose that α and $\beta = \alpha^x$ are distinct points with $x \in G$, and write $Q(\beta) = Q^x$.*

    (i) *α and β are perpendicular if and only if, for one (and hence all) $z \in Q$ with $[Q(\beta), Q(\beta)^z] \neq 1$, Q acts on $\{Q(\beta)^r \mid r \in R\}$, where R is the root group containing z.*

    (ii) *If α and β are perpendicular, then $[N_Q(Q(\beta)), N_{Q(\beta)}(Q)] = Q \cap Q(\beta) = R(\langle \alpha, \beta \rangle)$. Moreover, $|\alpha^{Q(\beta)}| = q$; and, if S is any root group inside Q not perpendicular to $Q \cap Q(\beta)$ within the orthogonal space Q, then S is transitive on the points of $\langle \alpha, \beta \rangle - \{\alpha\}$.*

Proof. (i) If α and β are not perpendicular, then β^Q is the set of points not perpendicular to α, and hence cannot coincide with β^R. Conversely, if α and β are perpendicular, then Q fixes the line $\langle \alpha, \beta \rangle$ and hence acts on the stated set of size q (corresponding to the points $\neq \alpha$ of $\langle \alpha, \beta \rangle$).

(ii) Note that $[N_Q(Q(\beta)), N_{Q(\beta)}(Q)] \subseteq Q \cap Q(\beta) = R(\langle \alpha, \beta \rangle)$. If L and M are totally singular lines such that $\alpha \in L \subseteq \beta^{\perp}$ and $\beta \in M \subseteq \alpha^{\perp}$ but $M \not\subseteq L^{\perp}$, then $[R(L), R(M)] = R(\langle \alpha, \beta \rangle)$ by **4.1.2**. $\square$

Pairs of root elements are not as well–behaved as in **4.1.2**. We will need the following property that resembles **4.1.2**(iv) but does not appear to be in print (cf. Lemma 4.2(ii)):

Lemma 4.8   (i) *Assume that $p > 2$. Let t and r be root elements but not long root elements such that t and t^r are distinct and commute, and r and r^t commute. Then $[t, r]$ is a long root element. Moreover, $[V, t]^\circ \in [V, r]^\perp$ and $[V, r]^\circ \in [V, t]^\perp$.*

(ii) *If t and r are long root elements such that t and t^r are distinct and commute, and r and r^t commute, then $[t, r]$ is a long root element.*

Proof. (i) Let $L = [V, t]$, $\alpha = [V, t]^\circ$, $M = [V, r]$ and $\beta = [V, r]^\circ$ (cf. Lemma 4.2(ii)). We claim that $\alpha \in M^\perp$ and $\beta \in L^\perp$. By symmetry, it suffices to assume that $\alpha \notin M^\perp$ and derive a contradiction.

Since t and t^r commute, t fixes L^r as well as L, hence also $L \cap L^r$. However, t moves all points of L other than α (this is where we use the assumption $p > 2$, since $(w, w) = 2\varphi(w) \neq 0$ in (4.1)); and $\alpha \notin L \cap L^r$, since otherwise r would send the unique singular point α of L to that of L^r and hence fix α, whereas $\alpha \notin M^\perp = C_V(r)$. Thus, $L \cap L^r = 0$. Since t fixes L^r and $L = [V, t]$, it follows from (4.1) that $L^r \subseteq L^\perp$.

In particular, $\alpha^\perp$ contains both L and L^r. Since $M = [V, r]$, we have $L^r \subseteq \langle L, M \rangle$. Since $\langle L, L^r \rangle$ is a 4–space it must be $\langle L, M \rangle$. Then $L \cap M = 0$, so that $\langle L, M \rangle = \langle L, L^r \rangle \subseteq \alpha^\perp$, which contradicts $\alpha \notin M^\perp$.

Thus, $\alpha \in M^\perp$ and $\beta \in L^\perp$ as claimed. Then t^r induces 1 on $\alpha^\perp / \alpha$, and hence so does $[t, r]$. Similarly, $[t, r]$ induces 1 on $\beta^\perp / \beta$. It follows that $[t, r]$ lies in the long root group determined by $\langle \alpha, \beta \rangle$.

(ii) This follows from **4.1.2**. □

Remark. Part (i) is false if $p = 2$.

Lemma 4.9 *Let t be as in* Lemma 4.8. *If $d \geq 8$ then the probability that a conjugate r of t behaves as in that lemma is $\geq (q - 2)/2q^3 \geq 1/6q^2$.*

Proof. (i) We only need to deal with points α, β and 2–spaces L, M behaving as in the preceding proof. A singular point β lies in $L^\perp - L$ with probability $\geq 1/2q^2$. Given β, a 2–space M such that $\beta \in M \in L^G$ lies in $\langle \alpha, \beta \rangle^\perp / \langle \alpha, \beta \rangle$, but not in $\langle L, \beta \rangle^\perp / \langle \alpha, \beta \rangle$, with probability at least $1 - 2/q$.

(ii) This is similar but easier. □

$O(1)$ **generation.**   Finally, we note the elementary fact that G and $(G_{\alpha\gamma})'$ can be generated by $O(1)$ matrices, which are easily written down using any basis of V or $\langle \alpha, \gamma \rangle^\perp$. If $d > 4$ then $(G_\alpha)'$ acts irreducibly on Q, and hence can be generated by the generators of $(G_{\alpha\gamma})'$ together with any nontrivial element of Q.

4.1.4   Probabilistic generation

Theorem 4.10   (a) [Ka2] *If $d \geq 5$ then each irreducible subgroup H of $G = \Omega^\varepsilon(d,q)$ generated by an H–conjugacy class of long root groups is one of the following:*

(i) $\Omega^\varepsilon(d,q)$,

(ii) $\mathrm{SU}(2n,q)$ *inside* $\Omega^+(4n,q)$ *or* $\mathrm{SU}(2n+1,q)'$ *inside* $\Omega^-(4n+2,q)$,

(iii) *a central extension of $\Omega(7,q)$ inside $\Omega^+(8,q)$ (obtained from the usual subgroup $\Omega(7,q)$ by conjugating with a triality outer automorphism of $\mathrm{P}\Omega^+(8,q)$),*

(iv) $G_2(q)$ *inside* $\mathrm{P}\Omega(7,q)$,

(v) *a subgroup of $\mathrm{P}\Omega^\pm(2n,2)$ preserving a decomposition of the $2n$–space as an orthogonal direct sum of anisotropic 2–subspaces, or*

(vi) $3\mathrm{P}\Omega^-(6,3)2$ *inside* $\mathrm{P}\Omega^+(12,2)$.

(b) [Co, (3.8)] *If H is a subgroup of G generated by long root elements, and if $O_p(H) = 1$, then H is the product of pairwise commuting subgroups each of which is transitive on the set of long root groups it meets nontrivially.*

For future reference we note that, in (ii–vi), there is no subgroup $\Omega^-(6,q)$ acting in the natural manner on a 6–space.

We now turn to probabilistic generation, as in **3.1.4**. We will not attempt to obtain estimates that are at all precise.

Lemma 4.11   (i) *If $d \geq 7$ then, with probability $> (1-1/q)^4/8 \geq 1/2^7$, three nontrivial long root elements generate a subgroup $J \cong \mathrm{SL}(3,q)$ acting reducibly on the nonsingular 6–space $[V,J]$, having two 3–dimensional constituents on this 6–space while inducing 1 on $[V,J]^\perp$.*

(ii) *If $d \geq 9$ then, with probability $\geq (1-1/q)^6/2^7 \geq 1/2^{13}$, four nontrivial long root elements generate a subgroup J inducing $\Omega^-(8,q)$ on the nonsingular 8–space $[V,J]$ and 1 on $[V,J]^\perp$. (The same estimate holds if G is $\Omega^-(8,q)$.)*

Proof. For $i = 1, 2, 3$ or 4, let t_i be randomly and independently chosen long root elements, let J denote the group they generate, and let $L_i = [V,t_i]$. In each case we will gradually build up the subspace spanned by the L_i.

(i) We can choose any L_1. Then $V_4^+ := \langle L_1, L_2 \rangle$ is a nonsingular 4–space with probability $q^{d-3}q^{d-4}/|L_1^G| > (1 - 1/q)/2$.

Next consider any 6–space V_6^+ of Witt index 3 containing V_4^+; the number of these is the same as the number of hyperbolic 2–spaces in $V_4^{+\perp}$. Choose L_3 inside V_6^+. In order to calculate it is easier at this stage to temporarily switch from the group $\Omega^+(6,q)$ to the group $\mathrm{PSL}(4,q)$. Then t_1 and t_2 become transvections, and we are choosing a third transvection t_3 within this group. By (the proof of) Lemma 3.7(i), $\langle t_1, t_2, t_3 \rangle$ fixes a point and a plane of V_6^+, and is isomorphic to $\mathrm{SL}(3,q)$, with probability $> (1 - 1/q)^3$; when this occurs, $V_6^+ = \langle V_4^+, L_3 \rangle$.

Thus, the desired probability is more than $\{(\#V_6^+ \text{ containing } V_4^+)(1 - 1/q)/2\} \{(\#L_3 \text{ contained in } V_6^+)(1 - 1/q)^3\}/|L_3^G| \geq \{(1 - 1/q)^4/2\}/4$.

(ii) The desired probability is at least that for $V_6^+ := \langle L_1, L_2, L_3 \rangle$ to be as in (i), multiplied by the probability that L_4, t_4, $V_8^- := \langle V_6^+, L_4 \rangle$ and J are then of the desired sort. Here, the probability that L_4 and V_8^- behave correctly is at least $(\#V_8^- \text{ containing } V_6^+)\{(\#L_4 \text{ contained in } V_8^-)(1 - 1/q)/4\}/|L_4^G| \geq (1-1/q)^2/16$, since the probability is at least $(1-1/q)/4$ that a totally singular line L_4 of V_8^- has trivial intersection with V_6^+. For each such L_4 we have $J = \langle t_1, t_2, t_3, t_4 \rangle = \Omega(V_8^-)$ using Theorem 4.10, since J is irreducible and $\langle t_1, t_2, t_3 \rangle \cong \mathrm{SL}(3,q)$ is generated by long root groups.

This time it follows that the probability that $\langle t_1, t_2, t_3, t_4 \rangle$ behaves as required is $> \{(1 - 1/q)^4/8\}\{(1 - 1/q)^2/16\}$. $\square$

Lemma 4.12 (i) *Suppose that $d \geq 5$. If either $q \geq 8$ is even or $q \geq 17$ then, with probability $\geq 1/640$, for two elements g_1, g_2 of the same $\mathrm{ppd}^\#(p; e)-$ or $\mathrm{ppd}^\#(p; 2e)-$order > 4 such that each $[V, g_j]$ is a nonsingular 2–space, $\langle g_1, g_2 \rangle$ induces 1 on $[V, \langle g_1, g_2 \rangle]^\perp$ and $\Omega^+(4,q)$ on the nonsingular 4–space $[V, \langle g_1, g_2 \rangle]$.*

(ii) *Suppose that $d \geq 7$ and either $q \geq 4$ is even or $q \geq 17$. Then, with probability $> 1/2^{10}10$, for three G–conjugate elements g_1, g_2, g_3 of the same $\mathrm{ppd}^\#(p; e)-$order such that each $[V, g_j]$ is a nonsingular 2–space, $\langle g_1, g_2, g_3 \rangle$ induces $\Omega^+(6,q)$ on the nonsingular 6–space $[V, \langle g_1, g_2, g_3 \rangle]$ and 1 on $[V, \langle g_1, g_2, g_3 \rangle]^\perp$.*

(iii) *Suppose that $d \geq 9$ and $q = 2$. Then, with probability $> 1/8$, for two elements g_1, g_2 of order 5 such that each $[V, g_j]$ is a nonsingular 4–space, $\langle g_1, g_2 \rangle$ induces $\Omega^-(8,2)$ on the nonsingular 8–space $[V, \langle g_1, g_2 \rangle]$ and 1 on $[V, \langle g_1, g_2 \rangle]^\perp$.*

Proof. Let $U_i := [V, g_i]$.

(i) We begin by showing that, *with probability* $> 1/32$, $U_2 \nsubseteq U_1^\perp$ *and* $V_4^+ := \langle U_1, U_2 \rangle$ *is a nonsingular 4–space of Witt index 2 on which* $J := \langle g_1, g_2 \rangle$ *is irreducible.* Choose any U_1. Each such 4–space V_4^+ is determined as $V_4^+ := \langle U_1, V_2 \rangle$ for a unique $V_2 \in U_1^G \cap U_1^\perp$, where $|U_1^G \cap U_1^\perp| > q^{2d-8}/4$. For each such V_4^+ we choose $U_2 \subset V_4^+$ with $U_1 \cap U_2 = 0$ and $U_2 \neq U_1^\perp \cap V_4^+$; there are more than $q^4/4$ choices for U_2. Then the desired probability is greater than $(q^{2d-8}/4)(q^4/4)/|U_1^G| > (q^{2d-8}/4)(q^4/4)/2q^{2d-4} = 1/32$.

Now Corollary 3.9 produces the desired lower bound $(1/32)(1/20)$.

(ii) First suppose that $q \neq 4$. By (i), with probability $\geq 1/640$ we have $\langle g_1, g_2 \rangle = \Omega(V_4^+)$, where $V_4^+ := [V, \langle g_1, g_2 \rangle]$. The nonsingular 6–spaces V_6^+ containing V_4^+ and having Witt index 3 arise from the more than $q^{2d-12}/2$ hyperbolic 2–spaces in $V_4^{+\perp}$; and for each such V_6^+ we choose one of the more than $q^8/4$ members U_3 of U_1^G lying in V_6^+ and having 0 intersection with both V_4^+ and $V_4^{+\perp}$. The probability that a member U_3 of U_1^G produces a nonsingular 6–space V_6^+ in this manner is greater than $(q^{2d-12}/2)(q^8/4)/|U_1^G| > (q^{2d-12}/4)(q^8/4)/2q^{2d-4} = 1/16$.

Thus, with probability $> (1/640)(1/16)$, $J := \langle g_1, g_2, g_3 \rangle$ induces an irreducible subgroup of $\Omega(V_6^+)$ containing $\Omega(V_4^+)$. Using Theorem 3.6 we find that J induces $\Omega(V_6^+)$, as required.

Now consider the case $q = 4$. As in (i), $V_4^- := \langle U_1, U_2 \rangle$ is a nonsingular 4–space of Witt index 1 with probability $\geq 1/32$. Then $\langle g_1, g_2 \rangle$ lies in $H = \Omega(V_4^-) \cong \mathrm{PSL}(2, 16)$. The maximal subgroups of H generated by two conjugates of $\langle g_1 \rangle$ are conjugate to $K = \mathrm{PSL}(2, 4)$, and there are $|N_H(\langle g_1 \rangle){:}N_K(\langle g_1 \rangle)| = 4 + 1$ of them. The union of these subgroups contains fewer than $5 \cdot 10$ subgroups of order 3. Thus, $\langle g_1, g_2 \rangle = H$ with probability $\geq 1 - 5 \cdot 10/|g_1^H| = 1 - 50/17 \cdot 8 > 0.6$.

As above, $V_6^+ := \langle U_1, U_2, U_3 \rangle$ is a nonsingular 6–space of Witt index 3 with probability $\geq 1/32$. Thus, with probability $> (0.6)(1/32)$, $\langle g_1, g_2, g_3 \rangle$ is an irreducible subgroup of $\Omega^+(6, 4) \cong \mathrm{PSL}(4, 4)$ containing $\Omega^-(4, 4) \cong \mathrm{PSL}(2, 16)$, and hence is $\Omega^+(6, 4)$.

(iii) Here U_i is a 4–space of Witt index 1. This time, *with probability* $\geq 1/5$, $\langle U_1, U_2 \rangle$ *is a nonsingular 8–space of Witt index 3 on which* $J = \langle g_1, g_2 \rangle$ *is irreducible.* Namely, for fixed U_1 the desired probability is at least $(\#4\text{–spaces } V_4^+ \text{ of Witt index 2 inside } U_1^\perp)(\#U_2 \subset \langle U_1, V_4^+ \rangle \text{ with } U_1 \cap U_2 = 0)/|U_1^G|$. Let $d = 2n$. Then there are $2^{4n-16}(2^{n-2} \mp 1)(2^{2n-6} - 1)(2^{n-4} \pm 1)/2 \cdot (2^2 - 1)^2$ of these V_4^+, and $|U_1^G| = 2^{4n-8}(2^n \mp 1)(2^{2n-2} - 1)(2^{n-2} \mp 1)/2 \cdot (2^2 - 1)(2^2 + 1)$; the number of indicated subspaces U_2 is obtained from the total number $17 \cdot 21 \cdot 2^7$

of members U_1^G lying in $\langle U_1, V_4^+ \rangle$ by subtracting at most $5 \cdot (17 \cdot 21 \cdot 2^7) 5/17 \cdot 7 + 10 \cdot (17 \cdot 21 \cdot 2^7) 5/17 \cdot 7$ (corresponding to the numbers of these subspaces on a singular or a nonsingular point of V_4^+). It follows that the desired probability is at least $(1/2)(1/2^8) \cdot (1/2^8) \cdot (2^7 \cdot 229) > 1/5$.

Now we can focus on irreducible subgroups $J = \langle g_1, g_2 \rangle$. All proper irreducible subgroups L of $\Omega^-(8,2)$ of order divisible by 5 are conjugate, and are isomorphic to $\Omega^-(4,4) \cdot 2$ (e.g., using [CCNPW]). The probability that g_1 and g_2 lie in one of these conjugates is at most $|G:L| \binom{17}{2}^2 / |\langle g_1 \rangle^{\Omega^-(8,2)}|^2 \leq 1/8 \cdot 3 \cdot 7$.

Thus, the probability in (iii) is at least $(1/5)(1 - 1/8 \cdot 3 \cdot 7) > 1/8$. $\square$

4.1.5 Irreducible and half-irreducible elements

$O^-(2n, q)$ contains elements of order $q^n + 1$, and each is irreducible [KL, p. 122]. In view of Lemma 2.5, $\Omega^-(2n,q)$ *has at least* $|\Omega^-(2n,q)|/4n$ *elements of* $\mathrm{ppd}^\#(p; 2en)$*-order*, and then this order is necessarily a factor of $q^n + 1$.

$O^+(2n, q)$ contains elements of order $q^n - 1$ that split V as $V = U_1 \oplus U_2$ for totally singular n-spaces U_1, U_2 on each of which the element is irreducible. If n is odd then each element of $\mathrm{ppd}^\#(p; en)$-order fixes two such n-spaces but no nonsingular n-space; as in Lemma 2.5, *there are at least* $|\Omega^+(2n,q)|/4n$ *such elements of* $\Omega^+(2n,q)$.

We need one property of such an element g of $\mathrm{ppd}^\#(p; en)$-order when n is odd: *for any vector v in neither U_1 nor U_2 we have* $V = \langle vg^i \mid 0 \leq i \leq 2n-1 \rangle$. For, we can write $v = u_1 + u_2$ with $u_i \in U_i$. If the indicated vectors are linearly dependent then there is a polynomial f of least degree such that $vf(g) = 0$ and $0 < \deg f < 2n$, and then f divides the minimal polynomial m of g. On the other hand, the minimal polynomial m_i of g on U_i is irreducible of degree n, and $m_1 \neq m_2$. (Otherwise, $m = m_1$, so that $\langle v'g^i \mid 0 \leq i \leq n-1 \rangle$ is a g-invariant n-space for any nonsingular vector v', which is impossible since n is odd.) Thus, $m = m_1 m_2$. Moreover, $f \neq m_1, m_2$ since $v \notin U_1, U_2$. It follows that $\deg f \geq 2n$, which is not the case.

4.2 Long root groups

We now begin our algorithm for the orthogonal case of Theorems 1.1 and 1.1'. The cases $d \leq 6$ are essentially handled in other sections for isomorphic copies of these groups (however, see **4.6.1** and **4.6.3**). The cases $d = 7, 8$ are postponed to **4.6.2**. Consequently, we will *assume that $d \geq 9$*.

4.2.1   Finding long root elements

We will find a long root element t of G by considering several cases. Except in **1** below we make *at most $\lceil 8qd\ln(4d)\rceil$ tests of randomly chosen elements τ of G* and check their orders; t will be a power of τ in **2** and **3**. We will also arrange that t satisfies additional conditions for later use; for example, t and a suitable power τ' of τ fix a common singular point α of the target vector space V.

Each case is followed by a Note, *stating properties of the various elements and subgroups of G if G is assumed to be acting projectively on V.* These notes are for purposes of explanations and later proofs: they are nonalgorithmic.

Case 1. $p \neq 2$ **and** $q < 17$: Choose up to $\lceil 4q^2 d\ln(4d)\rceil$ elements τ of G in order to find one of $p \cdot \mathrm{ppd}^{\#}(p; e(2n-4))$–order, where $d = 2n-1$ or $2n$.

Then $a := \tau^z$ *is a root element*, where $z := q^{n-2}+1$. For, we have chosen $|\tau|$ so that τ must split V as $V = \mathrm{C}_V(\tau^p) \perp \mathrm{C}_V(\tau^p)^\perp$, where $\mathrm{C}_V(\tau^p)$ is a nonsingular 3– or 4–dimensional subspace while τ^p induces on $\mathrm{C}_V(\tau^p)^\perp$ an irreducible element of order dividing z. All p–elements of the orthogonal group on $\mathrm{C}_V(\tau^p)$ are root elements for $\mathrm{C}_V(\tau^p)$ and thus also for V. By **4.1.5** and Lemma 2.5, a single choice for τ behaves as required with probability $\geq (1/q^2)(1/2(2n-2)) \geq 1/2q^2 d$. Thus, at least one of our $\lceil 4q^2 d\ln(4d)\rceil$ choices has the desired order with probability $> 1 - 1/16d^2$.

Now choose up to $\lceil 12q^2 \ln(4d)\rceil$ conjugates b of a, and for each test whether $[a,b] \neq 1$ and $[a, a^b] = 1 = [b, b^a]$. Such an element b determines *a long root element* $t := [a, b]$ by Lemma 4.8 (this is where the assumption $p \neq 2$ is critical). By Lemma 4.9 the probability is $\geq 1/6q^2$ that a single conjugate b will behave as in Lemma 4.8; hence, the probability is $> 1 - 1/16d^2$ that at least one of our elements b is as desired.

Note: $t \in \mathrm{O}_p(G_\alpha)$ for the unique singular point $\alpha = [V, a] \cap [V, t]$ of $[V, t]$ fixed by $\tau' := \tau^{q+1}$. (The exponent is needed since τ can induce an element of order dividing $p(q+1)$ on a 4–space when $q+1 | q^{n-2}+1$.) Also, $\tau'' := \tau^{p(q+1)}$ acts irreducibly on the nonsingular $2n-4$–space $[V, \tau'']$.

Time: $O(d\log d[\xi + \mu d^2])$ since q is bounded. The above procedure works for all odd q, but we wish to avoid a factor q^2 in our timing. On the other hand, we needed to single out the present case in order to avoid minor difficulties that would arise later for small odd q.

In the remainder of **4.2.1** we may assume that $q \geq 17$ if q is odd.

Case 2. $G/\mathrm{Z}(G) \cong \Omega^+(2n, q)$ **with** $n \geq 5$ **odd**: Find $\tau \in G$ of $p \cdot \mathrm{ppd}^{\#}(p; 2e) \cdot \mathrm{ppd}^{\#}(p; e(n-2))$–order; we also require that $8 \big| |\tau|$ if q is a

Mersenne prime, and $|\tau| = 18\mathrm{ppd}^{\#}(2; e(n-2))$ if $q = 8$. (Recall that all of these requirements for $|\tau|$ mean that $|\tau|$ is divisible by the primes, or 4, or 9 in the present situation, involved in the definition of these $\mathrm{ppd}^{\#}$; cf. **2.4**.)

Using **4.1.5** we find that an element of G has such an order with probability $\geq 2(1/q)(1/4) \cdot 1/2(2n-4) \geq 1/4qd$, and $t := \tau^{wz}$ *is a long root element*. For, such an element τ must split V as $V = V_4^{+} \perp V_{2n-4}^{+}$, inducing elements of order dividing $pw := p(q+1)$ on V_4^{+} and $z := q^{n-2} - 1$ on V_{2n-4}^{+} since $(w, z) \leq 2$. Since τ^z and τ^{pw} commute, τ^z induces 1 on V_{2n-4}^{+} and acts within $\Omega(V_4^{+}) \cong \mathrm{SL}(2, q) \circ \mathrm{SL}(2, q)$ as an element of $p \cdot \mathrm{ppd}^{\#}(p; 2e)$–order $\geq 3p$. Then $t = \tau^{wz}$ has order p and lies in one of the $\mathrm{SL}(2, q)$ factors, and thus is indeed a long root element.

The probability that at least one of up to $\lceil 8qd\ln(4d) \rceil$ choices for τ behaves as desired is $> 1 - 1/16d^2$.

Note: There are $q + 1$ singular points α fixed by $\tau' := \tau^w$ lying in the support $[V, t] \subset V_4^{+}$ of t, and $t \in O_p(G_\alpha)$ for each such α. Also, $\tau'' := \tau^{pw}$ has two irreducible constituents on $[V, \tau''] = V_{2n-4}^{+}$, each of which is totally singular.

Case 3. $G/Z(G) \cong \Omega^{-}(2n, q)$ **with** $n \geq 6$ **even:** Find $\tau \in G$ of $p \cdot \mathrm{ppd}^{\#}(p; 2e) \cdot \mathrm{ppd}^{\#}(p; e(2n-4))$–order; we also require that $8\|\,|\tau|$ if q is a Mersenne prime, and that $|\tau| = 18\mathrm{ppd}^{\#}(2; e(2n-4))$ if $q = 8$. Such an element τ must split V as $V = V_4^{+} \perp V_{2n-4}^{-}$, inducing elements of order dividing $pw := p(q+1)$ on V_4^{+} and $z := q^{n-2} + 1$ on V_{2n-4}^{-} since $(w, z) \leq 2$; and hence τ occurs with probability $\geq 2(1/q)(1/4) \cdot 1/2(2n-4) \geq 1/4qd$. As in **2**, for such an element $t := \tau^{wz}$ *is a long root element*.

Note: See the Note in **2**, except that this time $\tau'' := \tau^{pw}$ acts irreducibly on the nonsingular $2n - 4$–space $[V, \tau'']$.

Case 4. $G/Z(G) \cong \Omega^{+}(2n, q)$ **with** $n \geq 6$ **even and** $q \geq 8$: Find $\tau \in G$ of $\mathrm{ppd}^{\#}(p; e) \cdot \mathrm{ppd}^{\#}(p; 4e) \cdot \mathrm{ppd}^{\#}(p; e(2n-6))$–order; we also require that $8\|\,|\tau|$ if either $q = p$ is a Fermat prime or $q = p^2$ with p a Mersenne prime. Such an element τ must split V as $V = V_2^{+} \perp V_4^{-} \perp V_{2n-6}^{-}$, inducing elements of order dividing $h := q - 1$ on V_2^{+}, $w := q^2 + 1$ on V_4^{-} and $z := q^{n-3} + 1$ on V_{2n-6}^{-} since h, w and z pairwise have $\gcd \leq 2$; and hence τ occurs with probability $\geq (1/4)(1/8) \cdot 1/2(2n-6) \geq 1/64d$. Then $a := \tau^{wz/(2,q-1)}$ induces an element of $\mathrm{ppd}^{\#}(p; e)$–order ≥ 3 on V_2^{+} and 1 on $V_2^{+\perp}$.

Now choose up to $\lceil 2^{11}\ln(4d) \rceil$ conjugates b of a, and for each test whether $\langle a, b \rangle \cong \Omega^{+}(4, q)$ using **3.6.2** (cf. **3.2.3**). When this occurs use this isomorphism to find one of the $\mathrm{SL}(2, q)$ factors of $\langle a, b \rangle$, let T be one of its Sylow

p–subgroups normalized by a, and let $1 \neq t \in T$. Then t *is a long root element.*

By Lemma 4.12(i), for a single choice of b the group $\langle a, b \rangle$ is as required with probability $\geq 1/640$, and then **3.6.2** succeeds with probability $\geq 3/4$, hence both succeed with probability $\geq (1/640)(3/4) > 1/2^{10}$. Thus, the probability is $> 1 - 1/16d^2$ that at least one of our choices for b produces $\langle a, b \rangle \cong \Omega^+(4, q)$ and that this is proven using **3.6.2**, in which case we obtain an isomorphism $\langle a, b \rangle \to \Omega^+(4, q)$ and the required group T.

Note: There are four possibilities for T normalized by a: two in each $\mathrm{SL}(2, q)$ factor. Each of these fixes each singular point α of the nonsingular 2–space $[V, a] < [V, \langle a, b \rangle]$, and $t \in O_2(\langle a, b \rangle_\alpha) \leq O_2(G_\alpha)$. Moreover, $\tau' := \tau$ fixes $\alpha = [V, t] \cap [V, a]$ but no other singular point of $[V, t]$. Finally, $\tau'' := \tau^{hw}$ acts irreducibly on the nonsingular $2n - 6$–space $[V, \tau'']$.

Case 5. $G/Z(G) \cong \Omega(2n+1, q)$ **with** $n \geq 3$ **and odd** $q \geq 17$, **or** $\Omega^-(2n, q)$ **with** $n \geq 5$ **odd and** $q \geq 8$: Find $\tau \in G$ of $\mathrm{ppd}^\#(p; e) \cdot \mathrm{ppd}^\#(p; e(2n - 2))$–order; where once again we require that $8 \big| |\tau|$ if either $q = p$ is a Fermat prime or $q = p^2$ with p a Mersenne prime. Such an element τ must split V as $V = V_* \perp V_2^+ \perp V_{2n-2}^-$ with $\dim V_* \leq 1$, inducing elements of order dividing $h := q - 1$ on V_2^+ and $z := q^{n-1} + 1$ on V_{2n-2}^- since $(h, z) \leq 2$; and hence τ occurs with probability $\geq (1/4) \cdot 1/2(2n - 2) \geq 1/8d$. Then $a := \tau^z$ induces an element of $\mathrm{ppd}^\#(p; e)$–order ≥ 3 on V_2^+ and 1 on $V_2^{+\perp}$. Now proceed as in the preceding case.

Note: See the Note in **4**, except that here $\tau'' := \tau$ acts irreducibly on the nonsingular $2n - 2$–space $[V, \tau'']$.

Case 6. $G \cong \Omega^+(2n, 4)$ **with** $n \geq 6$ **even:** Find $\tau \in G$ of $51\mathrm{ppd}^\#(2; 2(2n-6))$–order. Such an element τ must split V as $V = V_2^+ \perp V_4^- \perp V_{2n-6}^-$, inducing an element of order 3 on V_2^+, of order 17 on V_4^- and of order dividing $z := 4^{n-3} + 1$ on V_{2n-6}^- since 3, 17 and z are pairwise relatively prime; and hence τ occurs with probability $\geq (1/4)(1/8) \cdot 1/2(2n - 6) \geq 1/64d$. Then $a := \tau^{17z}$ induces an element of order 3 on V_2^+ and 1 on $V_2^{+\perp}$.

Now choose up to $\lceil 2^{11} 10 \ln(4d) \rceil$ pairs b_1, b_2 of conjugates of a, and for each of them list $\langle a, b_1, b_2 \rangle$ and test whether $\langle a, b_1, b_2 \rangle \cong \Omega^+(6, 4)$ using brute force. When this occurs use brute force to find the center T of a Sylow 2–subgroup normalized by a, and let $1 \neq t \in T$. Then t *is a long root element.*

By Lemma 4.12(ii), for a single choice of b_1, b_2 the group $\langle a, b_1, b_2 \rangle$ is as required with probability $\geq 1/2^{10} 10$. Then the probability is $> 1 - 1/16d^2$ that at least one of our choices for b_1, b_2 produces $\langle a, b_1, b_2 \rangle \cong \Omega^+(6, 4)$.

Note: $t \in O_2(\langle a, b \rangle_\alpha) \leq O_2(G_\alpha)$; $\tau' := \tau^3$ is 1 on $[V, a]$ and hence fixes

$\alpha = [V,a] \cap [V,t]$ but no other point of $[V,t]$. Finally, $\tau'' := \tau^{hw}$ acts irreducibly on the nonsingular $2n-6$–space $[V,\tau'']$.

Case 7. $G \cong \Omega^-(2n,4)$ **with** $n \geq 5$ **odd:** Find $\tau \in G$ of $3\text{ppd}^\#(2; 2(2n-2))$–order. Such an element τ must split V as $V = V_2^+ \perp V_{2n-2}^-$, inducing an element of order 3 on V_2^+ and of order dividing $z := 4^{n-1} + 1$ on V_{2n-2}^- since $(3,z) = 1$; and hence τ occurs with probability $> (1/4) \cdot 1/2(2n-2) \geq 1/8d$. Then $a := \tau^z$ induces an element of order 3 on V_2^+ and 1 on $V_2^{+\perp}$. Now proceed as in the preceding case.

Note: See the Note in **6**, except that here $\tau'' := \tau$ acts irreducibly on the nonsingular $2n-2$–space $[V, \tau'']$.

Case 8. $G \cong \Omega^+(2n,2)$ **with** $n \geq 6$ **even:** Find $\tau \in G$ of $5\text{ppd}^\#(2; 2n-6)$–order. Such an element τ must split V as $V = V_2^+ \perp V_4^- \perp V_{2n-6}^-$, inducing an element of order 5 on V_4^- and of order dividing $z := 2^{n-3} + 1$ on V_{2n-6}^- since $(5,z) = 1$; and hence τ occurs with with probability $\geq (1/5) \cdot 1/2(2n-6) \geq 1/2d$. Then $a := \tau^z$ induces an element of order 5 on V_4^- and 1 on $V_4^{-\perp}$.

Now choose up to $\lceil 16 \ln(4d) \rceil$ conjugates b of a, list $\langle a,b \rangle$, and test whether it is $\Omega^-(8,2)$. If so, there is a unique subgroup $F \cong \Omega^-(4,2)$ of $\langle a,b \rangle$ that contains a (and centralizes a 4–space of the 8–space underlying $\Omega^-(8,2)$). Let α be a singular point of the 4–space underlying F, and let t be a nontrivial long root element of $O_2(\langle a,b \rangle_\alpha)$ that does not commute with τ^5. (Note that $|C_{O_2(\langle a,b \rangle_\alpha)}(\tau^5)| \leq 2^4$ while $|O_2(\langle a,b \rangle_\alpha)| = 2^6$ using **4.1.3**.)

By Lemma 4.12(iii), for a single choice of b the group $\langle a,b \rangle$ is as required with probability $> 1/8$; hence the probability is $> 1 - 1/16d^2$ that $\langle a,b \rangle \cong \Omega^-(8,2)$ for at least one of our choices b.

Note: $t \in O_2(\langle a,b \rangle_\alpha) \leq O_2(G_\alpha)$; $\tau' := \tau^5$ is 1 on $[V,a] = V_4^-$ and hence fixes $\alpha = [V,a] \cap [V,t]$ but no other singular point of $[V,t]$ (since t and τ' do not commute). Also, $\tau'' := \tau^5$ acts irreducibly on the nonsingular $2n-6$–space $[V, \tau'']$.

Case 9. $G \cong \Omega^-(2n,2)$ **with** $n \geq 5$ **odd:** Find $\tau \in G$ of $5\text{ppd}^\#(2; n-2)$–order. Such an element τ must split V as $V = V_4^- \perp V_{2n-4}^+$, inducing an element of order 5 on V_4^- and of order dividing $z := 2^{n-2} - 1$ on V_{2n-4}^+ since $(5,z) = 1$; and hence τ occurs with probability $\geq (1/5) \cdot 1/2(2n-4) \geq 1/10d$. Then $a := \tau^z$ induces an element of order 5 on V_4^- and 1 on $V_4^{-\perp}$ and hence we can proceed as in **8**.

Note: See the Note in **8**, except that this time $\tau'' := \tau^5$ has two irreducible constituents on $[V, \tau''] = V_{2n-4}^+$, each of which is totally singular.

Reliability: $> 1 - 1/8d^2$.

Time: Las Vegas $O(d \log d[\xi + \mu d^2] + qd \log d[\xi + \mu e d^2 \log q] + \log d[\xi qe + \mu q \log^2 q])$ since the test in **3.6.2** takes $O(\xi qe + \mu q \log^2 q)$ time.

4.2.2   Finding the subgroups J and J_k

We now construct $O(\log d)$ subgroups $J_k \cong \Omega^-(8,q)$ (recall that $d \geq 9$). We need this many such subgroups for the probability estimate in Lemma 4.13.

Choose up to $\lceil 2^{20} \ln(4d) \rceil$ triples $f_1, f_2, f_3 \in G$, and for each choice let $J := \langle t, t^{f_1}, t^{f_2}, t^{f_3} \rangle$. For a single choice, $J \cong \Omega^-(8,q)$ with probability $\geq 1/2^{13}$ by Lemma 4.11(ii). If indeed $J \cong \Omega^-(8,q)$ then the algorithm in **4.6.2** (when combined with **7.2.2**) verifies this fact with probability > 0.99 (alternatively, see **3.2.3**). By Lemma 2.8 with the parameters $r = 0.99/2^{13}$, $\varepsilon = 1 - 1/3.96$ and $t = \lceil 2^{16} \log(4d) \rceil$, with probability at least $1 - 1/16d^2$ we obtain at least $\lceil 2^5 \ln(4d) \rceil$ subgroups $J \cong \Omega^-(8,q)$.

For each of these subgroups J_k, $1 \leq k \leq \lceil 2^5 \ln(4d) \rceil$, our use of **4.6.2** yields an 8–dimensional vector space V_{J_k}, an isomorphism $\lambda_{J_k} \colon J_k \to \Omega(V_{J_k})$ and a generating set $\mathcal{S}^*_{J_k}$ consisting of root elements. For these k find the following:

(i) The long root group R containing t;

(ii) $Q_{8k} := O_p(\text{stabilizer } (J_k \lambda_{J_k})_{\alpha_k} \lambda_{J_k}^{-1}$ of a point α_k of $V_{J_k})$ with $t \in Q_{8k}$, so that $|Q_{8k}| = q^6$ (*the choice of α_k will be discussed below*); a generating set $\mathcal{S}^*_{Q_{8k}}$ for Q_{8k} consisting of long root elements of J_k including t, such that each long root group that meets $\mathcal{S}^*_{Q_{8k}}$ is generated by its intersection with $\mathcal{S}^*_{Q_{8k}}$;

(iii) $G_{8k} = ((J_k \lambda_{J_k})_{\alpha_k})' \lambda_{J_k}^{-1}$, the subgroup of $((J_k \lambda_{J_k})_{\alpha_k}) \lambda_{J_k}^{-1}$ generated by its long root groups; and a subgroup $D_k \cong \Omega^-(6,q)$ such that $G_{8k} = Q_{8k} \rtimes D_k$ and $D_k \lambda_{J_k}$ fixes a point γ_k of V_{J_k} not perpendicular to α_k;

(iv) $j(\gamma_k) \in J_k$, where $j(\gamma_k) \lambda_{J_k}$ moves α_k to γ_k, together with a straight-line program of length $O(\log q)$ from $\mathcal{S}^*_{J_k}$ to $j(\gamma_k)$ (*we will abbreviate $J := J_1$, $\alpha := \alpha_1$, $\gamma := \gamma_1$, $D := D_1$ and $j(\gamma) := j(\gamma_1)$*);

(v) $s \in J$ of order $(q-1)/(2, q-1)$, normalizing R and $R^{j(\gamma)}$ and centralizing D; we can find a straight-line program of length $O(\log q)$ from $\mathcal{S}^*_J$ to any given power of s; and

(vi) s^+ such that $s^+ \lambda_J$ induces an element of order $q - 1$ on the 1–space α; we can find a straight-line program of length $O(\log q)$ from $\mathcal{S}^*_J$ to any given power of s^+ (when q is even use $s^+ = s$).

We have already used **4.6.2**. It is straightforward to write the required matrices and pull them back to J_k using $\lambda_{J_k}^{-1}$, exactly as in the proof of Corollary 3.20. Note that $j(\gamma_k)$ exists by (4.1), but it cannot be a long root element since α and γ are not perpendicular.

Choice of α_k: Presently all that we know about α_k is that it lies in $[V_{J_k}, t]$ (which we think of as "$[V, t]$"). We need to choose the α_k in such a way that they *all* are "fixed" by τ' and correspond to the same point of V. In **4-9** and part of **1** there is exactly one point of "$[V, t]$" fixed by τ'; in the remaining cases all points of "$[V, t]$" are fixed.

For each k find $g'_k \in J_k \lambda_{J_k}$ of order $q+1$ with $\langle g'_k \rangle$ transitive on the points of the totally singular line $[V_{J_k}, t\lambda_{J_k}]$. Find $g_k := g'_k \lambda_{J_k}^{-1}$. We will use a power of g_k to conjugate all of the elements and subgroups in (i-vi) to others that interact correctly.

Make initial choices of the points $\alpha_k \in V_{J_k}$ lying in $[V_{J_k}, t\lambda_{J_k}]$, and find the corresponding groups Q_{8k} (cf. (ii)).

1. *Choice of α_k in those cases in which τ' does not centralize t.* In each of these cases we must replace α_k by a certain unique point of each $[V_{J_k}, t\lambda_{J_k}]$ such that all of these points coincide within V if we think of all V_{J_k} as subspaces of V. Fix k.

   Find i such that $0 \le i \le q$ and $[Q_{8k}^{g_k^i}, Q_{8k}^{g_k^i \tau'}] = 1$; replace Q_{8k} by $Q_{8k}^{g_k^i}$ and hence α_k by $\alpha_k^{(g'_k)^i}$.

   Correctness: In view of the hypotheses of this section, there is an epimorphism $\lambda^{\#}: G \to \mathrm{P}\Omega(V)$. Then $J_k \lambda^{\#}$ acts on the 8–space $V_k := [V, J_k \lambda^{\#}]$, inducing 1 on $V_k^{\perp}$. Note that $Q_8(i) = Q_{8k}^{g_k^i} \lambda^{\#} < G$ uniquely determines a singular point $\alpha(i)$ of V: the radical of $[V, Q_8(i)]$. Let α^* denote the point of $[V, t\lambda^{\#}]$ fixed by $\bar{\tau} = \tau'\lambda^{\#}$. It suffices to work within groups, replacing $Q_{8k} = Q_8(0)$ by a suitable conjugate $Q_8(i)$, and hence not having to deal with λ_k at all.

   There is some i for which $\alpha(i) = \alpha^*$, and then $[Q_8(i), Q_8(i)^{\bar{\tau}}] = 1$ by **4.1.3**. Conversely, suppose that $[Q_8(i), Q_8(i)^{\bar{\tau}}] = 1$. Each totally singular line $\ne \langle \alpha^*, \alpha(i) \rangle = [V, t\lambda^{\#}]$ of V_k through $\alpha(i)$ is disjoint from all members of a set of totally singular lines of $V_k^{\bar{\tau}}$ through $\alpha(i)^{\bar{\tau}}$ that span $\alpha(i)^{\bar{\tau}\perp} \cap V_k^{\bar{\tau}}$. Thus, from **4.1.2** it follows that $\alpha(i)^{\perp} \cap V_k$ and $\alpha(i)^{\bar{\tau}\perp} \cap V_k^{\bar{\tau}}$ are perpendicular. Both of these sets contain $\alpha^* = \alpha^{*\bar{\tau}}$. Thus, $\alpha^{*\perp} \cap V_k \supseteq \alpha(i)^{\perp} \cap V_k$, while V_k is nonsingular, so that $\alpha^* = \alpha(i)$, as desired.

2. *Choice of all α_k in those cases in which τ' centralizes t.* This time we choose α_1 arbitrarily in $[V_{J_1}, t]$, and then we must replace the remaining α_k so they will coincide with it. Consider any $k \geq 2$.

Find i such that $0 \leq i \leq q$ and $[Q_{8k}^{g(i)_k}, Q_{81}] = 1$; Q_{8k} by $Q_{8k}^{g_k^i}$ and hence α_k by $\alpha_k^{(g_k')^i}$.

Correctness: As above, $J_1\lambda^\#$ and $J_k\lambda^\#$ act on the 8–spaces $[V, J_1\lambda^\#]$ and $[V, J_k\lambda^\#]$, respectively. Moreover, $Q_{8k}(i) = Q_{8k}^{g_k^i}\lambda^\#$ determines the radical $\alpha(i)$ of $[V, Q_{8k}(i)]$ while $Q_1^* = Q_{81}\lambda^\#$ determines the radical α^* of $[V, Q_1^*]$.

If $\alpha(i) = \alpha^*$ then $[Q_{8k}(i), Q_1^*] = 1$ by **4.1.3**. Suppose that $[Q_{8k}(i), Q_1^*] = 1$ but $\alpha(i) \neq \alpha^*$. Each totally singular line $\neq [V, t\lambda^\#]$ through α^* of the 7–space $[V, Q_1^*]$ is disjoint from all members of a set of totally singular lines of $[V, Q_{8k}(i)]$ through $\alpha(i)$ that span $[V, Q_{8k}(i)]$. Thus, from **4.1.2** it follows that $[V, Q_1^*]$ and $[V, Q_{8k}(i)]$ are perpendicular. Both of these subspaces contain $[V, t\lambda^\#]$, so the 2–space $[V, t\lambda^\#]$ is contained in the radical $[V, Q_{8k}(i)]^\perp \cap [V, Q_{8k}(i)] = \alpha(i)$ of $[V, Q_{8k}(i)]$. This contradiction proves that $\alpha(i) = \alpha^*$.

From now on we may assume that *all Q_{8k} lie in the same group 'Q'* (corresponding to some point of V) *and that τ' normalizes 'Q'*.

Increasing the sets $\mathcal{S}_{J_k}^*$: We assume that these are increased to include $\mathcal{S}_{Q_{8k}}^*$ and a generating set of D_k consisting of long root elements.

Reliability: $> 1 - 1/16d^2$ to find J_k, λ_{J_k} and $\mathcal{S}_{J_k}^*$ behaving as required.

Time: Las Vegas $O(\log d[\xi qe + \mu q \log^2 q])$: $O(\log d[\xi qe + \mu q \log^2 q])$ to choose the elements f_1, f_2, f_3 and test the isomorphisms $J_k \cong \Omega^-(8, q)$, then $O(\log d[\mu qe])$ to use Proposition 4.23 to find all of the elements $t\lambda_{J_k}$, $O(\log d[\mu e \log q])$ to use Proposition 4.22 to find the $O(e)$ elements and subgroups in (i-vi), and $O(\log d[q \cdot \mu e^2])$ to adjust the points α_k.

Generating 'Q': By construction, each subgroup J_k is a random member of the conjugacy class of naturally embedded $\Omega^-(8, q)$ subgroups of G, subject only to the condition that $t \in J_k$. Then, within the orthogonal space 'Q' (cf. **4.1.3**), each Q_{8k} is a random nonsingular subspace having dimension 6, Witt index 2 and containing t. These $\lceil 32\ln(4d)\rceil$ subspaces generate 'Q' with probability bounded away from 0. However, we require a more precise probability estimate using the element τ'' defined in **4.2.1**:

Lemma 4.13 'Q' $= \langle Q_{8k}, [\text{'}Q\text{'}, \tau''] \mid 1 \le k \le \lceil 32 \ln(4d) \rceil \rangle$ *holds with probability* $> 1 - 1/16d^2$.

Proof. In each case listed in **4.2.1**, $Y = [\text{'}Q\text{'}, \tau'']$ is a nonsingular subspace of the orthogonal $d - 2$–space 'Q' such that $\dim Y = 2n - 2$, $2n - 4$ or $2n - 6$. Moreover, τ'' either acts irreducibly on Y or splits Y as the direct sum of two totally singular invariant subspaces of odd dimension $(\dim Y)/2$ on each of which τ'' acts irreducibly.

Now we consider the various cases. Fix k. We consider the probability that Q_{8k} lies in a hyperplane of 'Q' that contains Y. We note the following straightforward count: (#) *If U is a hyperplane of* 'Q'*, then the probability is* $< 2/q^5$ *that Q_{8k} contains t and lies in U.*

5,7 with d even in **5**: Here $Y =$ 'Q': there is no such hyperplane.

1,5 with d odd in **5**: If $d = 2n+1$ then, by (#), Q_{8k} lies in the $2n-2$–space Y with probability $< 2/q^5$. If $d = 2n$ then Q_{8k} lies in one of the $q+1$ hyperplanes of 'Q' containing the $2n - 2$–space Y with probability $< (q + 1)2/q^5$.

2,3,9: Once again Q_{8k} lies in one of the $q+1$ hyperplanes of 'Q' containing the $2n - 4$–space Y with probability $< (q + 1)2/q^5$.

4,6,8: This time Q_{8k} lies in one of the $(q^2 + 1)(q + 1)$ hyperplanes of 'Q' containing the $2n - 6$–space Y with probability $< (q^2 + 1)(q + 1)2/q^5 \le 15/16$.

Thus, 'Q' $= \langle Q_{8k}, [\text{'}Q\text{'}, \tau''] \rangle$ with probability $\ge 1/16$. Consequently, this equality holds for at least one of our $\lceil 32 \ln(4d) \rceil$ choices J_k with probability $> 1 - 1/16d^2$. $\square$

Remarks. We have used the group $\Omega^-(8, q)$ in order to avoid minor difficulties with $\Omega^+(8, q)$ caused by the triality automorphism. We did not use $\Omega^+(6, q)$ in order to have available the observation following the statement of Theorem 4.10. We could not use $\Omega^-(6, q)$: in (iii) we need long root elements, whereas none exist in $\Omega^-(4, q)$.

4.3   Finding $Q = Q(\alpha)$, $Q(\gamma)$, G'_α, L and $\mathcal{S}^*$

4.3.1 Q, $Q(\gamma)$ and G'_α

With τ' as defined in **4.2.1**, let

$$Q = Q(\alpha) := \langle Q_{8k}^{\tau'p^i} \mid 0 \leq i \leq d,\ 1 \leq k \leq \lceil 32\ln(4d)\rceil \rangle$$
$$G'_\alpha := \langle G_{8k}^{\tau'p^i} \mid 0 \leq i \leq d,\ 1 \leq k \leq \lceil 32\ln(4d)\rceil \rangle$$
$$Q(\gamma) := Q^{j(\gamma)}.$$

We relate these to the groups 'Q' and '(G_α)''= $\mathrm{N}_G($'Q'$)'$ (cf. **4.1.3**):

Lemma 4.14 *With probability* $> 1 - 1/16d^2$, $Q = $ 'Q' *and* $G'_\alpha = \mathrm{N}_G(Q)'$.

Proof. By Lemma 4.13 we may assume that 'Q' is spanned by the Q_{8k} and $Y = ['Q', \tau'']$. Thus, $Q = $ 'Q' if $Q \geq Y$.

   Recall that τ'' either acts irreducibly on Y or has two totally singular invariant subspaces of odd dimension $(\dim Y)/2$. If τ'' acts irreducibly then Q contains Y by Lemma 2.7. The cases in which τ'' is not irreducible are **2** and **9**. In each of these, for each 6–space Q_{8k} we have $\dim(Q_{8k} \cap Y) \geq 4$ since $\dim Y = 2n - 4$, so that Q_{8k} contains a nonsingular vector of Y; since $\dim Y \equiv 2 \pmod 4$, it follows that Q contains Y by the last paragraph of **4.1.5**.

   We now turn to $H = G'_\alpha/Q$, which we may view as generated by subgroups $D_k^{\tau'p^i}$, $1 \leq k \leq \lceil 32\ln(4d)\rceil$, each of which is generated by long root groups of the orthogonal group $\Omega(Q)$.

   We first claim that $O_p(H) = 1$. For otherwise, H would leave invariant a nonzero totally singular subspace W. For $0 \leq i \leq d$ and $1 \leq k \leq \lceil 32\ln(4d)\rceil$, W is $D_k^{\tau'p^i}$–invariant, so that $W \subseteq (Q_{8k}^{\tau'p^i})^\perp$. Consequently, W is in the intersection of these subspaces, which we have already shown is $Q^\perp = 0$. This contradiction proves our claim.

   By Theorem 4.10(b), it follows that H is the product of commuting subgroups H_j, $j = 1, \ldots, l$, each of which is generated by the H_j–class consisting of the long root groups it contains, and hence also by an H_j–class of conjugates of various subgroups D_k. Let $Y_j = [Q, H_j]$. If $0 \leq i \leq d$ and $1 \leq k \leq \lceil 32\ln(4d)\rceil$, then $D_k^{\tau'p^i}$ either lies in H_j or commutes with H_j, so that it leaves Y_j invariant and hence either $Y_j \supseteq Q_{8k}^{\tau'p^i}$ or $Y_j \subseteq Q_{8k}^{\tau'p^i\perp}$. Since Q is spanned by these subspaces $Q_{8k}^{\tau'p^i}$, it follows that Y_j is spanned by some of these subspaces. Consequently, the Y_j are pairwise perpendicular.

Now $Q = Y_1 \perp \cdots \perp Y_l$, where $l \leq (d-2)/\dim Q_{8k} = (d-2)/6$. In particular, $l = 1$ if $d < 14$, and $6 \leq \dim Y_j \leq d - 2 - 6$ if $l \geq 2$.

In order to show that $l = 1$ in general, we need to use the behavior of τ'^p, even though that transformation is not (yet) known to permute the subspaces Y_j. Assume that $l \geq 2$. Since τ'^p does not leave invariant any nonsingular subspace of Q of dimension between 5 and $d - 7$ (cf. **4.2.1**), it must move each Y_j and hence does not normalize any H_j. Consider the graph on the G–conjugates of D lying in H, joining two conjugates if they do not commute; each connected component generates a subgroup H_j. We may assume that $Q_8^{\tau'^{pd}} \leq H_l$. Then the τ'^p–images of those connected components lying in $H_1 \cup \cdots \cup H_{l-1}$ must lie in connected components of our graph. Thus, we can renumber the H_j so that $H_j^{\tau'^p} \leq H_{j+1}$ for $1 \leq j \leq d - 1$. By considering τ'^{-p} we see that $\tau'^p \colon H_1 \mapsto H_2 \mapsto \cdots \mapsto H_l$ and hence $\tau'^p \colon Y_1 \mapsto Y_2 \mapsto \cdots \mapsto Y_l$. Then $Y_l^{\tau'^p}$ must lie in $\langle Y_1, \ldots, Y_{l-1} \rangle^{\tau'^p \perp} = Y_1$. Thus, τ'^p permutes the Y_j cyclically. However, τ'^p has a power θ whose order is a prime $r = \mathrm{ppd}^{\#}(p; k)$ with $k \geq e(d-2)/3$ in each case in **4.2.1** (note that $2n - 6 \geq (2n-2)/3$ since $d \geq 14$). Then $k | r - 1$. Since θ acts nontrivially on $\{Y_1, \ldots, Y_l\}$, this produces the contradiction $(d-2)/3 \leq k < r \leq l \leq (d-2)/6$.

Thus, $Q = Y_1$. It follows that H acts irreducibly on Q. For, if W is a proper H–invariant subspace then, as above, $Q_{8k}^{\tau'^{pi}} \subseteq W \cup W^{\perp}$ for all i, k. Then each $D_{8k}^{\tau'^{pi}}$ fixes W and $W^{\perp}$, and hence the long root groups lying in $H_1 = H$ cannot all be conjugate.

By Theorem 4.10, G'_{α} induces $\Omega(Q)$ on Q except, perhaps, if $\Omega(Q)$ is $\Omega^+(8, q)$ and G'_{α} is $\Omega(7, q)$ acting irreducibly. However, it is not hard to check that the latter group does not contain a subgroup $\Omega^-(6, q)$ acting in the usual manner (although it does contain a central extension of $\mathrm{PSU}(4, q) \cong \mathrm{P}\Omega^-(6, q)$ acting irreducibly on the 8–space). $\square$

Root groups: Q_{8k} and Q are generated by long root groups. We assume that generating sets for all of these long root groups are stored so that for each generator we can find the root group containing it (cf. **4.2.2**(iii)).

4.3.2   Effective transitivity of Q; the complement L

Our present generating set for Q has size $O(ed \log d)$. In **4.4.3**(4) we will obtain a generating set of $O(ed)$ long root elements, at which point the running times for the algorithms in this subsection can have all of their $\log d$ factors deleted, except for Corollary 4.18 (which is used for **4.4.3**(4)).

Points: Conjugates of Q will be called *points*.

Equality testing: Two points are equal if and only if they commute; however, there is a faster test. Let $g, h \in Q$ be generators lying in different root groups (cf. the end of **4.3.1**); then $Q^x = Q^y$ if and only if $[g^x, Q^y] = 1 = [h^x, Q^y]$, which are tested using one generator for each of the $O(d \log d)$ long root groups generating Q^y. (For, this condition occurs if and only if Q^y fixes $[V, g^x] \cap [V, h^x] = \alpha^x$.) **Time:** $O(\mu d \log d)$.

Perpendicular points: Two distinct points Q^x and Q^y are called *perpendicular* if and only if the following holds for some (and hence all) $z \in Q^x$ with $Q^{yz} \neq Q^y$: Q^x acts on $\{Q^{yu} \mid u \in Z\}$, where Z is the root group containing z. Note that this imitates Lemma 4.7(i). The following lemma tests for perpendicularity; it is unfortunate that it involves listing the elements of subgroups of order q.

Lemma 4.15 *There is a deterministic algorithm that*

(i) *in $O(\mu q[e + d \log d])$ time tests whether or not two given distinct points Q^x and Q^y are perpendicular; and*

(ii) *if they are perpendicular, in $O(\mu q e[d \log d]^2)$ time finds $O(e)$ generators for $Q^x \cap Q^y$.*

Proof. (i) Let z be a generator of Q^x, so that z is a long root element. If $Q^y = Q^{yz}$ then Q^x and Q^y are perpendicular by (4.1). Assume that $Q^y \neq Q^{yz}$. Find the long root group $Z < Q^x$ containing z (cf. the end of **4.3.1**). List Z using Lemma 2.1. Then Q^x and Q^y are perpendicular if and only if, for each generator g of Q^x, there is some $u \in Z$ such that $Q^{ygu} = Q^y$. In order to speed this test up we use only one choice of g, namely a long root element in $Q^x - Z$.

Correctness: If α^x and α^y are not perpendicular then $|(\alpha^y)^{\langle Z, g \rangle}| > q$ since Q^x is regular on the points not perpendicular to Q^x.

Time: Dominated by listing Z and finding u.

(ii) Test $O(d \log d)$ elements z lying in different long root groups generating Q^x in order to find one that does not normalize Q^y. Again let Z be the long root group containing z. Then the regularity of Z on $\{Q^{yu} \mid u \in Z\}$ implies that $N_{Q^x}(Q^y)$ is generated by the elements gu as g ranges over the generators of Q^x. We will not need this group, but instead we will use $N_{Q^y}(Q^x)$, which is obtained in the same manner.

Now let $z^* \in Q^x - Z$ be another generator of Q^x, and let Z^* be the long root group containing it. If z^* normalizes Q^y let $A := Z^*$.

If z^* does not normalize Q^y then, as above, for each of the $O(e)$ generators $g \in Z^*$ find $u(g) \in Z$ such that $gu(g)$ normalizes Q^y; these elements $gu(g)$ generate a subgroup A of Q^x.

In either case we have obtained a root group $A < Q^x$ that normalizes Q^y. Let a be a generator of A. Test all generators h of $N_{Q^y}(Q^x)$ until one is found having $[a, h] \neq 1$. List $[A, h]$; this group is $Q^x \cap Q^y$ and has $O(e)$ generators.

Correctness: We have described generators of $N_{Q^x}(Q^y)$: they are "Schreier generators" for this subgroup. We obtained a root group $A = Z^*$ or $(ZZ^*)_{\alpha^y}$ (though this is not necessarily a long root group).

Since A normalizes Q^y it fixes α^x and α^y, but $[V, A]^\perp \not\supseteq \langle \alpha^x, \alpha^y \rangle^\perp$. Thus, there are lines of $\langle \alpha^x, \alpha^y \rangle^\perp$ on α^y that are not in $\langle \alpha^x, \alpha^y \rangle^\perp \cap [V, A]^\perp$. Correspondingly, there is some generator h of $N_{Q^y}(Q^x)$ such that $[V, h]$ is not in $\langle \alpha^x, \alpha^y \rangle^\perp \cap [V, A]^\perp$. Note that $[A, h] \neq 1$ (as otherwise h would fix $[V, A]$, whereas $[V, h] \cap [V, A] = 0$ and $[V, h] \not\subseteq [V, A]^\perp$). Now $[A, h]$ lies in both Q^x and Q^y; it is a root group, and hence its order is that of $Q^x \cap Q^y$.

Time: $O(\mu q e [d \log d]^2)$, dominated by finding $N_{Q^y}(Q^x)$. $\square$

In one circumstance we can deal with perpendicularity much faster than in the preceding lemma:

Lemma 4.16 *In deterministic $O(\mu[qe + d \log d])$ time, given a long root element u not normalizing Q^x, a point Q^w can be found containing u and perpendicular to Q^x.*

Proof. Test at most one element per long root group in order to find a generator $a \in Q^x$ such that $[[a, u], u] \neq 1$; let $A < Q^x$ be the long root group containing a (cf. **4.3.1**). List A^u. Find $b \in A^u$ such that $[u, a^b] = 1$. Output $Q^w := Q^{xb}$.

Correctness: First we show that a exists. For any generator a of Q^x, if $[[a, u], u] = 1$ then $[V, a] \cap [V, u]^\perp \neq 0$ by **4.1.2**; here $\alpha^x \in [V, a]$ but $\alpha^x \notin [V, a] \cap [V, u]^\perp$ since we have assumed that u does not fix α^x. The 2-spaces $[V, a]$ span $\alpha^{x\perp}$. If $[V, a] \cap [V, u]^\perp \neq 0$ for each generator a (or even just one generator per long root group), then the 1-spaces $[V, a] \cap [V, u]^\perp$ span at least a hyperplane of $\alpha^{x\perp}$ lying in $[V, u]^\perp$. Dimensions force $[V, u]^\perp$ to be that hyperplane, whereas $[V, u]^\perp \not\subseteq \alpha^{x\perp}$.

Thus, a exists, and $J = \langle A, u \rangle \cong SL(2, q)$ by **4.1.2**. Then A^u is a Sylow subgroup of J containing neither A nor u, and hence has an element b conjugating a into the Sylow p-subgroup of J containing u. Moreover, J fixes

$\langle \alpha^x, \alpha^{x\perp} \cap [V, u] \rangle$ (by **4.1.2**, since this line meets both $[V, a]$ and $[V, u]$ non-trivially), so that b sends α^x to $\alpha^{x\perp} \cap [V, u]$, as required.

Time: Dominated by finding a and listing A^u using Lemma 2.1. $\square$

Lemma 4.17 *In deterministic $O(\mu q e [d \log d]^2)$ time, given two points Q^x, Q^y not perpendicular to Q, an element of Q can be found conjugating Q^x to Q^y.*

Proof. If Q^x and Q^y are distinct and perpendicular, find $1 \neq w \in Q^x \cap Q^y$ and $Q \cap Q^w$ using Lemma 4.15(ii). Test the q elements $v \in Q \cap Q^w$ in order to find one such that $Q^{xv} = Q^y$.

If Q^x and Q^y are not perpendicular, for each generator r of Q (taking only one from any long root group) use Lemma 4.16 to find a point $Q^{r'}$ containing r and perpendicular to Q^x; find a generator r such that $Q^{r'}$ is not perpendicular to Q^y. Use Lemma 4.15 to find $1 \neq s \in Q^x \cap Q^{r'}$. Use Lemma 4.16 to find a point $Q^{x'}$ containing s and perpendicular to Q^y. The preceding paragraph produces elements of Q conjugating Q^x to $Q^{x'}$ and $Q^{x'}$ to Q^y.

Correctness: If α^x and α^y are distinct and perpendicular, then $Q^x \cap Q^y$ is the long root group $R(\langle \alpha^x, \alpha^y \rangle)$ by Lemma 4.7(ii). By (4.1), α and α^w are perpendicular; they are distinct since $\langle \alpha^x, \alpha^y \rangle = [V, w] \not\subseteq \alpha^\perp$. Then $Q \cap Q^w$ is the long root group $R(\langle \alpha, \alpha^w \rangle)$ by Lemma 4.7(ii), and $\langle \alpha^x, \alpha^y \rangle \cap \langle \alpha, \alpha^w \rangle$ is a point δ by (4.1). Here $R(\langle \alpha, \alpha^w \rangle)$ induces all transvections of $\langle \alpha, \delta, \alpha^x \rangle$ with axis $\langle \alpha, \delta \rangle$ and center δ, and hence is transitive on the points not in $\alpha^\perp$ of $\langle \alpha^x, \alpha^y \rangle$ by (4.1). Thus, $v \in Q \cap Q^w$ exists as stated.

When α^x and α^y are not perpendicular we found points $\alpha^{r'}$ spanning $\alpha^\perp \cap \alpha^{x\perp}$; one of these must not be in $\alpha^{y\perp}$. Then $\alpha^{x'}$ is not perpendicular to α since $\alpha^{r'}$ is the point of $[V, r] = \langle \alpha^x, \alpha^{r'} \rangle$ lying in $\alpha^\perp$, where $\alpha^{r'}$ is not perpendicular to α^y, and hence $\alpha^{x'} \notin \alpha^\perp$.

Time: $O(\mu q e [d \log d]^2)$. Namely, if Q^x and Q^y are perpendicular, in $O(\mu q e [d \log d]^2)$ time find $Q^x \cap Q^y$ using Lemma 4.15(ii), then find and list $Q \cap Q^w$, and test all $v \in Q \cap Q^w$, in $O(\mu q e [d \log d]^2 + \mu q e + q \cdot \mu d \log d)$ time using Lemmas 4.15(ii) and 2.1. If Q^x and Q^y are not perpendicular, we applied Lemmas 4.15(i) and 4.16 to $O(d \log d)$ choices of r, and then used Lemmas 4.15(ii) and 4.16 once again, before reverting to the previous case. $\square$

Remark. By (4.4) the element of Q found in the lemma is *unique*, and there is a complement to Q in G'_α. One can be found as in the proof of Corollary 3.12:

Corollary 4.18 *In deterministic $O(\mu q e [d \log d]^3)$ time a complement $L \cong G'_\alpha / Q \cong \Omega^\varepsilon(d - 2, q)$ to Q in G'_α can be found that normalizes the point $Q(\gamma)$.*

Proof. By **4.2.2**(iii) we have $Q_{8k} \leq Q$ and $Q_{8k}^{j(\gamma_k)} \leq Q^{j(\gamma_k)}$, where Q and $Q^{j(\gamma_k)}$ are not perpendicular since Q_{8k} and $Q_{8k}^{j(\gamma_k)}$ are not (e.g., use the test at the beginning of this section). Since Q_{8k} is contained in only one conjugate of Q, by **4.2.2**(iv) D_k normalizes both Q and $Q^{j(\gamma_k)}$. Use the lemma to modify each of the conjugates $D_k^{\tau'p_i}$ generating G'_α modulo Q to a conjugate $D_k^{\tau'p_i u}$, $u \in Q$, normalizing $Q(\gamma)$. Let L be generated by these conjugates $D_k^{\tau'p_i u}$. Then $G'_\alpha = QL$, and $Q \cap L = 1$ since L normalizes $Q(\gamma)$.

This requires $O(d \log d)$ uses of the algorithm in the lemma. $\square$

4.3.3   Recursion: $\mathcal{S}^*$ and λ_L

We will require that *our set $\mathcal{S}^*$ of generators of G consists entirely of root elements*. Recursively apply Theorem 1.1′ to L, obtaining *a $d-2$–dimensional orthogonal $\mathbb{F}$–space V_L, a set $\mathcal{S}_L^*$ of root elements generating L, and an isomorphism $\lambda_L \colon L \to \Omega(V_L) = \Omega^\varepsilon(d-2, q)$ defined on $\mathcal{S}_L^*$*.

Temporarily exclude the case $G/Z(G) \cong P\Omega^+(10, q)$. Then $G = \langle G'_\alpha, J \rangle$ is generated by the set

$$\mathcal{S}^* := \mathcal{S}^*(1) \cup \mathcal{S}^*(2) \cup \mathcal{S}^*(3),$$

where $\mathcal{S}^*(1) = \mathcal{S}_L^*$, $\mathcal{S}^*(2)$ is our present set of generators for Q (cf. **4.3.1**), and $\mathcal{S}^*(3)$ is the union of the $O(\log d)$ sets $\mathcal{S}_{J_k}^*$ (cf. **4.2.2**). At present $\mathcal{S}^*(2)$ has size $O(ed \log d)$. In Lemma 4.19 we will obtain a new generating set $\mathcal{S}^*(2)$ of size $O(ed)$, at which point $\mathcal{S}^*$ will have size $O(ed^2)$.

The case $G/Z(G) \cong P\Omega^+(10, q)$. Recursion is slightly less straightforward here: whereas **4.6.2**, or even just brute force, produces an 8–dimensional vector space for $L/Z(L)$, there are *three* vector spaces that might arise (permuted transitively by a triality automorphism of $L/Z(L)$). This situation is analogous to the one in **3.3.3**, where the possibility of two equally natural L–modules had to be considered. We eliminate two of the three vector spaces as follows.

Find commuting nontrivial long root elements $r_1, r_2 \in L\lambda_L$ such that $[V_L, r_1]$ and $[V_L, r_2]$ have nonzero intersection but are not perpendicular. Then $[[V_L, r_1], r_2] = [V_L, r_1] \cap [V_L, r_2]$ (cf. 4.1). Find $r_1 \lambda_L^{-1}$ and $r_2 \lambda_L^{-1}$ using Proposition 4.22. If $[[Q, r_1 \lambda_L^{-1}], r_2 \lambda_L^{-1}] \neq 1$ then λ_L is an isomorphism of the desired sort: the stabilizer of a nonzero singular vector of V_L is sent by λ_L^{-1} to the stabilizer of a nontrivial element of Q. For a choice of τ in **4.6.2** this will happen with probability $1/3$, assuming that **4.6.2** outputs an isomorphism.

Reliability: $> 1 - 1/4(10^2)$ after up to 20 repetitions of **4.6.2**: each repetition produces an isomorphism of the desired sort with probability $> (1 - 1/2 \cdot 8^2)/3$.

Time: $O(\xi + \mu e)$ in addition to the recursive call.

Now obtain V_L, λ_L and $\mathcal{S}^* = \mathcal{S}^*(1) \cup \mathcal{S}^*(2) \cup \mathcal{S}^*(3)$ as before.

Correctness: By **4.1.2**, there are two different ways two long root elements $r_1\lambda_L^{-1}$ and $r_2\lambda_L^{-1}$ can commute: $[Q, r_1\lambda_L^{-1}]$ and $[Q, r_2\lambda_L^{-1}]$ can either meet or be perpendicular; and, in the latter case, r_2 induces the identity on $[Q, r_2\lambda_L^{-1}]$. A triality outer automorphism of $L/Z(L)$ sends a pair of the first sort to one of the second sort.

Computing preimages recursively: As in **3.3.3**, we will need to be able to apply λ_L^{-1} to elements of $L\lambda_L$. This is the content of Proposition 4.22 (and of Theorems 1.1(iv) and 1.1'(iv)), with L in place of G: *in $O(\mu d^2 \log q)$ time $h\lambda_L^{-1}$ can be found for any given $h \in L\lambda$.*

Two generators: It is straightforward to write down 2 generators (or, if preferred, $O(1)$ generators) of $L\lambda_L$, chosen as a matter of convenience. Use Proposition 4.22 to pull them back to generators of L.

The isomorphism $\lambda^\#$:   Throughout this section we are assuming that there exists an isomorphism $\lambda^\#\colon G \to \Omega(V)$ or $\mathrm{P}\Omega(V)$. While the goal is to construct such an isomorphism, as in **3.3.3** we use its existence to motivate and prove correctness of subroutines in the algorithm. Recall the convention in **3.3.1** that, for example, "Q" denotes the matrix group studied in **4.1.3**. Clearly $\lambda^\#$ sends the nonperpendicular points $Q, Q(\gamma)$ to a nonperpendicular pair of points of V, which we may assume is α, γ in the notation of **4.1.3**. As in **3.4.4** we can use $\lambda^\#$ to see that λ_L extends to an isomorphism $\lambda^\bullet\colon G'_\alpha = QL \to$ "Q" "L" determined by a semilinear map $Q \to$ "Q". (Note that we dealt with the anomalous case $L \cong \Omega^+(8, q)$ above: even in that case the automorphism of L induced by $\lambda_L\lambda^{\#-1}$ arises from a semilinear map on Q rather than from a graph automorphism of $\mathrm{P}\Omega^+(8,q)$.) This time, $\lambda^\bullet|_Q$ also preserves the form on Q in the sense that, for some $k \in \mathbb{F}^*$ and $\sigma \in \mathrm{Aut}\mathbb{F}$, we have $\varphi''(u\lambda^\bullet) = k\varphi_Q(u)^\sigma$ for all $u \in Q$, where φ'' and φ_Q denote the quadratic forms on "Q" and Q, respectively.

By following $\lambda^\#$ with a field automorphism we see that there is an isomorphism $\lambda^\bullet$ of the desired sort such that the map $\lambda^\bullet|_Q$ is linear. In Lemma 4.20 we will find $\lambda^\bullet|_Q$. For now we note only that in the present case $\lambda^\bullet|_Q$ is uniquely determined up to a scalar transformation.

4.4   Deterministic algorithms for Q

All procedures in 4.4 and 4.5 will be deterministic.

4.4.1 Decomposition of Q

As promised earlier, we now reduce the size of our generating set for Q to $O(ed)$:

Lemma 4.19 *In deterministic $O(\mu d^2 \log q)$ time $O(e)$–generator subgroups $A_i \le Q$, $1 \le i \le d-2$, can be found such that the following hold:*

(i) $|A_i| = q$ *and* $Q = A_1 \times \cdots \times A_{d-2}$; *and*

(ii) *In deterministic $O(\mu d)$ time, any given $u \in Q$ can be expressed in the form $u = \prod_1^{d-2} a_i$ with $a_i \in A_i$.*

Proof. We have a representation λ_L of L on the space $V_L = \mathbb{F}^{d-2}$ equipped with a nonsingular quadratic form φ' and associated bilinear form $(\ ,\ )'$. We separate the cases $p > 2$ and $p = 2$, assuming that the forms on V_L arise in different manners for these different cases.

Case $p > 2$: We may assume that the form is obtained from a decomposition $V_L = W_1 \perp \cdots \perp W_{d-2}$ for nonsingular 1–spaces W_i with $W_1, \ldots, W_{d-3}$ isometric; for later reference, we rescale φ' so that W_1 contains a vector e_1 such that $\varphi'(e_1) = 1$. Let c' and g' be any elements of $L\lambda_L$ such that $c' \colon W_1 \mapsto W_2 \mapsto \cdots \mapsto W_{d-3}$ and $g' \colon W_1 \mapsto W_1, W_2 \mapsto W_3 \mapsto \cdots \mapsto W_{d-3}$.

If $2 \le i \le d-3$, let $x_i' \in L\lambda_L$ be -1 on W_1 and W_i, and 1 on the remaining W_j. Then $x_i' = x_2'^{g'^{i-1}}$ whenever $3 \le i \le d-3$. For $i = 1, 3$, let $x_{i,d-2}' \in L\lambda_L$ be -1 on both W_{d-2} and a 1–space $\overline{W}_i$ of $W_i + W_{i+1}$ isometric to W_{d-2}, and 1 on $(W_i + W_{i+1} + W_{d-2})^\perp$.

Use Proposition 4.22 to find $c := c'\lambda_L^{-1}$, $g := g'\lambda_L^{-1}$, $x_1 := x_1'\lambda_L^{-1}$ and $x_{i,d-2} := x_{i,d-2}'\lambda_L^{-1}$ for $i = 1, 3$. Let $x_i := x_2^{g^{i-1}}$ whenever $3 \le i \le d-3$.

Find generators r and r_{d-2} of Q such that

$$[[r, x_3], x_2] \ne 1 \quad \text{and} \quad [[r_{d-2}, x_{1,d-2}], x_{3,d-2}] \ne 1;$$

let $R < Q$ and $R_{d-2} < Q$ be the respective long root groups containing them (cf. **4.3.1**). Then

$$[[Q, x_3], x_2] = [[R, x_3], x_2] \quad \text{and} \quad [[Q, x_{1,d-2}], x_{3,d-2}] = [[R_{d-2}, x_{1,d-2}], x_{2,d-2}]$$

have e generators. Let

$$\begin{aligned}
A_1 &:= [[Q, x_3], x_2] \\
A_i &:= A_1^{c^{i-1}} \text{ for } 2 \le i \le d-3 \\
A_{d-2} &:= [[Q, x_{1,d-2}], x_{3,d-2}].
\end{aligned}$$

Then Q is the direct product of the subgroups A_i.

Given $u \in Q$ we wish to express $u = \prod_1^{d-2} a_i$, with $a_i \in A_i$. First find $a_1 := [[u, x_3], x_2]^{1/4}$, where $a^{1/2}$ can be calculated as $a^{(p+1)/2}$ for any $a \in A_i$. Then find $a_i := [va_1^{-1}, x_i]^{-1/2}$ for $2 \leq i \leq d-3$, as well as $a_{d-2} := u(\prod_1^{d-3} a_i)^{-1}$.

Correctness: We may assume that the standard basis vector e_i spans W_i for $1 \leq i \leq d-2$. If $v = \sum_i k_i e_i \in V_L$ then $4k_1 e_1 = [[v, x_3'], x_2'] \in W_1$ and $-2k_i e_i = [v - k_1 e_1, x_i']$ for $2 \leq i \leq d-3$ by elementary calculations. Similarly, $4k_{d-2} e_{d-2} = [[v, x_{1,d-2}'], x_{3,d-2}'] \in W_{d-2}$ since $x_{3,d-2}' = 1$ on $\overline{W}_1$.

At the end of **4.3.3** we saw that there is an isomorphism $\lambda^\bullet : QL \to \text{``}Q\text{''} \text{``}L\text{''}$ extending λ_L. Then, using (4.3) together with Lemma 4.6(a), we have $r(v)\lambda^{\bullet -1} = \prod_i r(k_i e_i)\lambda^{\bullet -1}$ with $r(4k_1 e_1)\lambda^{\bullet -1} = [[r(v)\lambda^{\bullet -1}, x_3], x_2] \in A_1$, and also $r(-2k_i e_i)\lambda^{\bullet -1} = [r(v - k_1 e_1)\lambda^{\bullet -1}, x_i] \in A_i$ for $2 \leq i \leq d-3$, and then $a_{d-2} \in A_{d-2}$, as asserted for an arbitrary element $u = r(v)\lambda^{\bullet -1}$ of Q.

Time: Dominated by five uses of Proposition 4.22.

Case $p = 2$: Let $d = 2m$. This time we may assume that the form is given so that $V_L = W_1 \perp \cdots \perp W_{m-1}$ with all W_i nonsingular 2–spaces such that $W_1, \ldots, W_{m-2}$ are isometric; we may assume that W_1 has a basis e_{1x}, e_{1y} such that $\varphi'(e_{1x}) = \varphi'(e_{1y}) = (e_{1x}, e_{1y})' = 1$.

Let c' be any element of $L\lambda_L$ sending $W_1 \mapsto \cdots \mapsto W_{m-2} \mapsto W_1$ and $W_{m-1} \mapsto W_{m-1}$. Let $e_{ix} = e_{1x}^{c'^{i-1}}$ and $e_{iy} = e_{1y}^{c'^{i-1}}$ for $2 \leq i \leq m-2$. Let $e_{m-1x} \in W_1$ with $\varphi'(e_{m-1x}) = 1$. If possible choose e_{m-1y} such that e_{m-1x}, e_{m-1y} is a basis of W_{m-1} and $(e_{m-1x}, e_{m-1y})' = 1$, $\varphi'(e_{m-1y}) \neq 0$; such a basis exists except when $q = 2$, in which case we let e_{m-1y} be any vector of $W_{m-1} - \langle e_{m-1x} \rangle$.

Since $L\lambda_L$ does not contain transvections, instead we use products of two transvections:

$$x_1' \text{ is } t_{21} := \begin{pmatrix} 1 & 0 \\ 1 & 1 \end{pmatrix} \text{ on } W_1 \text{ and a nontrivial transvection on } W_{m-1};$$

$$y_1' \text{ is } t_{12} := \begin{pmatrix} 1 & 1 \\ 0 & 1 \end{pmatrix} \text{ on } W_1 \text{ and a nontrivial transvection on } W_{m-1};$$

both are 1 on $(W_1 + W_{m-1})^\perp$.

In order to deal with the possibility that W_{m-1} is not isometric to W_1, when $q > 2$ we need additional elements of $L\lambda_L$. First let $f_0 \in W_1$ be such that $\varphi'(f_0) = \varphi'(e_{m-1y}) + 1$; write $\zeta := \varphi'(e_{m-1y})^{-1}$. Let $e_0 := e_{m-1x} + f_0$, so that the basis e_{m-1x}, e_0 of $\langle e_{m-1x}, e_0 \rangle$ is isometric to e_{1x}, e_{1y}. Now define the following six elements of $L\lambda_L$:

x'_{m-1} is t_{21} on W_1 and W_{m-1} and 1 on $(W_1 + W_{m-1})^\perp$

y'_{m-1} is t_{12} on W_1, $\begin{pmatrix} 1 & \zeta \\ 0 & 1 \end{pmatrix}$ on W_{m-1} and 1 on $(W_1 + W_{m-1})^\perp$

x'_0 is t_{21} on W_0 and W_2 and 1 on $(W_0 + W_2)^\perp$

y'_0 is t_{12} on W_0 and W_2 and 1 on $(W_0 + W_2)^\perp$

$x'_{2,1}$ is t_{21} on W_1 and W_2 and 1 on $(W_1 + W_2)^\perp$

$y'_{1,2}$ is t_{12} on W_1 and W_2 and 1 on $(W_1 + W_2)^\perp$.

(Note that x'_1 and y'_1 could have been defined to be x'_{m-1} and y'_{m-1} if $q > 2$.)

When $q = 2$ we also use elements $\bar{x}'_i$, $1 \leq i \leq 4$, inducing t_{12} on W_1 and four transvections of $W_{m-2} + W_{m-1}$ having linearly independent centers, and 1 on $(W_{m-2} + W_{m-1})^\perp$. (N.B.—We have had to slide around the fact that the orthogonal group on W_{m-1} may have order 2.)

Use Proposition 4.22 to find $c := c'\lambda_L^{-1}$, $x_1 := x'_1\lambda_L^{-1}$, $y_1 := y'_1\lambda_L^{-1}$, $x_{m-1} := x'_{m-1}\lambda_L^{-1}$, $y_{m-1} := x'_{m-1}\lambda_L^{-1}$, $x_{1,2} := x'_{1,2}\lambda_L^{-1}$, $y_{1,2} := y'_{1,2}\lambda_L^{-1}$, $x_0 := x'_0\lambda_L^{-1}$ and $\bar{x}'_i := \bar{x}'_i\lambda_L^{-1}$. Let $x_i := x_2^{c^{i-1}}$ and $y_i := y_2^{c^{i-1}}$ for $2 \leq i \leq m-2$.

Let

$$A_{1x} \quad := [[Q, y_1], x_1] \qquad A_{1y} \quad := [[Q, x_1], y_1]$$
$$A_{ix} \quad := A_{1x}^{c^{i-1}} \qquad\qquad A_{iy} \quad := A_{1y}^{c^{i-1}} \quad \text{for } 2 \leq i \leq m-2$$
$$A_{m-1x} := [[Q, y_{m-1}], x_{m-1}] \quad A_{m-1y} := [[Q, x_{m-1}], y_{m-1}]$$

except that, when $q = 2$, A_{m-1x} and A_{m-1y} are any distinct subgroups of $C_A(\langle x_{m-2}, y_{m-2} \rangle$ where $A := [Q, \{\bar{x}_1, \bar{x}_2, \bar{x}_3, \bar{x}_4\}]$ has order 16. Then Q is the direct product of all of the subgroups A_{ix} and A_{iy}.

Any given $u \in Q$ can be written $u = \prod_i^{m-1}(a_{ix}a_{iy})$, where

$$a_{ix} := [[v, y_i], x_i] \quad a_{iy} := [[v, x_i], y_i] \quad \text{for } 1 \leq i \leq m-2.$$

Finally, we deal with $u' := u \prod_1^{m-2}(a_{ix}a_{iy}) \in A_{m-1x}A_{m-1y}$ somewhat less directly: if $q = 2$ there are only four possibilities to consider for u'; when $q > 2$ find $u'' := [u', x_0]$, $a_{m-1y} := u''[[u'', x_{1,2}], y_{1,2}][[u'', y_{12}], x_{12}]$ and $a_{m-1x} := u'a_{m-1y}$.

The time requirement is the same as when $p > 2$.

Correctness: The calculations within V_L are similar to the earlier ones, and then $\lambda^\bullet$ yields the required a_{ix} and a_{iy}. For the part involving u'', note that if $v' := v + \sum_1^{m-2}(k_{ix} + k_{iy})$ and $v'' := [v', x'_0]$, then $k_{m-1y}e_{m-1y} := v'' + [[v'', x'_{1,2}], y'_{1,2}] + [[v'', y'_{1,2}], x'_{1,2}]$ and $k_{m-1x}e_{m-1x} := v' + k_{m-1y}e_{m-1y}$. $\square$

Remarks. In the two cases, $\langle e_i \rangle \lambda^{\bullet-1} = A_i$ and $\langle e_{ix} \rangle \lambda^{\bullet-1} = A_{ix}$, $\langle e_{iy} \rangle \lambda^{\bullet-1} = A_{iy}$ for each i.

For use in **4.6.2** we note that the above algorithms are also valid for $d = 7, 8$.

We used c and g only to cut the timing by a factor of d.

4.4.2   A field of endomorphisms of Q

We may assume that $q \neq p$. Let $s \in J$ be the element of order $(q-1)/(2, q-1)$ in **4.2.2**(v). When viewed in the target space V we see that it fixes the points α and γ and acts trivially on the orthogonal complement of their span in the 8–space V_J and hence in the entire d–space. By the remark following (4.5), the automorphism that s induces on Q by conjugation is scalar multiplication by an element of order $|s|$. Then $\mathrm{GF}(q)$ can be taken to be the $\mathrm{GF}(p)$–span of this automorphism; this is found using Lemma 2.4.

Time: $O(\mu q \log q)$.

Identification with $\mathbb{F}$.   We need to relate this field to the field $\mathbb{F}$ already available from L. For this purpose, let W_0 be any subspace of V_L of dimension 2 and Witt index 1; we may take this to be the subspace $W_1 + W_2$ or W_1 in **4.4.1**, depending on whether p is odd or 2. Write $s_0' := \begin{pmatrix} \rho^2 & 0 \\ 0 & \rho^{-2} \end{pmatrix} \in \Omega(W_0) \leq L\lambda_L$ in terms of a hyperbolic basis of W_0, where the generator $\rho \in \mathbb{F}^*$ is as in **2.3**. Find $s_0 := s_0' \lambda_L^{-1}$. Let u be any nontrivial element of Q, and replace s by a power so that $u^s = u^{s_0}$. Identify s_0 with ρ^2, and then identify ρ with the element of $\mathrm{GF}(q)$ whose square is s and whose minimal polynomial is the same as that of ρ (there are only two elements of $\mathrm{GF}(q)$ to test). **Time:** $O(d^2 \log q)$.

Time to apply a field element: We can use Lemma 2.4 to write any element of $\mathbb{F}^*$ as $\rho^{2i} + \rho^{2j}$ for some i, j; and then this element can be applied to any $u \in Q$, obtaining $(\rho^{2i} + \rho^{2j})u = u^{s^i} u^{s^j}$ in $O(\mu)$ time.

4.4.3 Q as an $\mathbb{F}$–space

(1) *Bases $\mathcal{B}$ and $\mathcal{B}_p$ of Q:* We construct an $\mathbb{F}$–basis $\mathcal{B}$ of Q using the notation in the proof of Lemma 4.19. First we find a suitable basis for V_L.

Case $p > 2$: We already have $e_1 \in W_1$ with $\varphi'(e_1) = 1$; we may assume that $e_i := e_1^{c'^{i-1}}$ for $2 \leq i \leq d-3$. Find a (new) vector e_{d-2} in $(W_{d-4} + W_{d-3}) - W_{d-3}$ such that $\varphi'(e_{d-2}) = 1$ unless $q = 3$, in which case find such a vector in $(W_{d-4} + W_{d-3} + W_{d-2}) - (W_{d-3} + W_{d-2})$. Also, find an element $l'(d-2) \in \Omega(V_L)$ such that $e_1^{l'(d-2)} = e_{d-2}$, and find $l(d-2) := l'(d-2)\lambda_L^{-1}$. For the basis $e_1, \ldots, e_{d-2}$ of V_L we now have, whenever $1 \leq i \leq d-2$, an element $l'(i) \in \Omega(V_L)$ such that $e_i = e_1^{l'(i)}$, as well as $l(i) = l'(i)\lambda_L^{-1} \in L$. Although this may no longer be an orthogonal basis of V_L, we can quickly pass between this basis and the one in **4.4.1** since only the behavior of e_{d-2} has been changed.

Case $p = 2$: Let $[[V_L, y_1'], x_1'] = \langle e_{1x} \rangle$ and $[[V_L, x_1'], y_1'] = \langle e_{1y} \rangle$ with $\varphi'(e_{1x}) = \varphi'(e_{1y}) = 1$; define e_{2x}, e_{2y}, e_{m-1x} and e_{m-1y} similarly. (Each of the indicated vectors is in the center of an orthogonal transvection and hence is nonsingular.) Let $e_{ix} := e_{2x}^{c'^{i-1}}$ and $e_{iy} := e_{2y}^{c'^{i-1}}$ for $3 \leq i \leq m-2$ (c' is defined in **4.4.1**). Let $r_{1y}', r_{2x}', r_{2y}', r_{m-1x}', r_{m-1y}' \in \Omega(V_L)$ send e_{1x} to $e_{1y}, e_{2x}, e_{2y}, e_{m-1x}, e_{m-1y}$, respectively, and find their images under λ_L^{-1}.

Reorder the basis $\{e_{ix}, e_{iy} \mid 1 \leq i \leq m-1\}$ of V_L in any way as $e_1, \ldots, e_{2m-2}$, and write $A_i := \langle e_i \rangle$ for $1 \leq i \leq 2m-2$. Once again, whenever $1 \leq j \leq 2m-2$ we can use the elements in the preceding two paragraphs to obtain $l'(i) \in \Omega(V_L)$ such that $e_i = e_1^{l'(i)}$, as well as $l(i) = l'(i)\lambda_L^{-1} \in L$.

For *any* p we can now mimic all of this in Q: we now let $1 \neq b_1 \in A_1$ and $b_i := b_1^{l(i)}$ for $1 \leq i \leq d-2$. Now $\mathcal{B} := \{b_i \mid 1 \leq i \leq d-2\}$ is an $\mathbb{F}$–basis of Q and $\mathcal{B}_p := \{b_i^{s^j} \mid 1 \leq i \leq d-2, \ 0 \leq j < e\}$ is a GF(p)–basis of Q.

Time: $O(\mu d^2 \log q)$.

(2) $\mathbb{F}$*–linear combinations*: Any given $u \in Q$ can be written as an $\mathbb{F}$–linear combination of $\mathcal{B}$, or of $\mathcal{B}_p$, or by means of a straight-line program of length $O(d \log q)$ from $\mathcal{B}_p$, by using Lemma 4.19(i) precisely as in **3.4.3**(4). **Time:** $O(\mu q d)$.

(3) *Another basis $\mathcal{Z}$*: We have another generating set for the $\mathbb{F}$–space Q using the union $\mathcal{Z}_p$ of generating sets of the groups $Q_{8k}^{\tau'^{pi}}$, $0 \leq i \leq d$, $1 \leq k \leq \lceil 32 \ln(4d) \rceil$. More precisely, by taking one long root element in each of the long root groups containing generators for these conjugates of Q_{8k} we obtain $O(d \log d)$ long root elements (cf. **4.2.2**(ii)); by (2) they can all be expressed as $\mathbb{F}$–linear combinations using $\mathcal{B}$ in $O(d \log d \cdot \mu q d)$ time. Use linear algebra to find an $\mathbb{F}$–basis $\mathcal{Z} = \{z_1, \ldots, z_{d-2}\}$ of Q using these $O(d \log d)$ linear combinations, so that $\mathcal{Z}$ consists of long root elements. Now replace $\mathcal{Z}_p$ by a subset consisting of of $e(d-2)$ members of $\mathcal{Z}_p$ such that, for each long root group A in Q, $\mathcal{Z}_p \cap A$ is either is empty or a basis of A. **Time:** $O(\mu q d^2 \log d + (d \log d)d^2)$, including the matrix algebra.

We have the transition matrix from $\mathcal{B}$ to $\mathcal{Z}$. Thus, any given $u \in Q$ can be expressed as an $\mathbb{F}$–linear combination of $\mathcal{Z}$ by using (2) followed by a matrix calculation. **Time:** $O(\mu q d + d^3)$.

(4) *Modify $\mathcal{S}^*(2)$*: As suggested in **4.3.3**, we now redefine $\mathcal{S}^*(2)$ as $\mathcal{S}^*(2) :=$ $\mathcal{B}_p \cup \mathcal{Z}_p$. Consequently, *whenever the algorithms in* **4.3.2** *are needed, we can now use* $\mathcal{Z}_p$ *to remove all factors* $\log d$ *from the timings within that section except for* Corollary 4.18. This is the main use of $\mathcal{Z}_p$. (The construction of $\mathcal{B}_p$ in Lemma 4.19 shows that it has no long root elements.)

We could also use $\mathcal{Z}_p$ in place of $\mathcal{B}_p$ throughout the remainder of this section, with relatively few changes. For example, in Lemma 4.19(i) we can use the transition matrix from $\mathcal{B}_p$ to $\mathcal{Z}_p$ in order to obtain another straight-line program from $\mathcal{Z}_p$ to any given element of Q, again in $O(\mu qd)$ time and again of length $O(d \log q)$.

(5) *Matrices $\tilde{g}$*: Using $\mathcal{B}$ and (2) we can find the matrix $\tilde{g}$ representing the linear transformation induced on the $\mathbb{F}$–space Q by any given $g \in \mathrm{N}_G(Q)$. **Time:** $O(d \cdot \mu qd)$.

4.4.4   The extension $\lambda^\bullet$

As in **3.4.4**, we now extend λ_L to G'_α. In **4.4.3**(1) we found a basis $e_1, \ldots, e_{d-2}$ of V_L and a basis $b_1, \ldots, b_{d-2}$ of Q (with the latter group written additively). First we identify V_L with $\mathbb{F}^{d-2}$ in such a way that $e_1, \ldots, e_{d-2}$ becomes the standard basis of $\mathbb{F}^{d-2}$. We then identify this space (in any manner) with the $d-2$–space $\langle \alpha, \gamma \rangle^\perp$ occurring in **4.1.3**. Recall that we have an isometry $r \colon \langle \alpha, \gamma \rangle^\perp \to$ "Q" in (4.3). Let "L" $\cong \Omega^\varepsilon(d-2, q)$ denote the centralizer of $\langle \alpha, \gamma \rangle$ in the orthogonal group "G" implicit in **4.1.3**.

Lemma 4.20 *In $O(\mu d^2 \log q)$ time λ_L can be extended to an isomorphism $\lambda^\bullet \colon G'_\alpha = QL \to$ "Q" "L" $\cong \mathbb{F}^{d-2} \rtimes \Omega^\varepsilon(d-2, q)$ with the following properties:*

(i) *if $b \in Q$ with $b = \sum_i k_i b_i$ (for $k_i \in \mathbb{F}$ found in $O(\mu qd)$ time), then $b\lambda^\bullet = r(\sum_i k_i e_i)$; $\lambda^\bullet$ is $\mathbb{F}$–linear and determines an L–invariant quadratic form φ_Q on Q such that $\lambda^\bullet|_Q$ becomes an isometry; and*

(ii) *$\tilde{l} = l\lambda^\bullet = l\lambda_L$ whenever $l \in L$.*

Proof. At the end of **4.3.3** we saw that λ_L extends to an isomorphism $\lambda^\bullet \colon QL \to$ "Q" "L" induced by a semilinear map $\lambda^\bullet|_Q$ that preserves the forms (up to a scalar and a field automorphism) and is uniquely determined up to a scalar transformation. The definition

$$b_i \lambda^\bullet := r(e_i) \text{ for } i \leq i \leq d-2.$$

was implicitly used in **4.4.1** and **4.4.3**(1). In fact, it is forced by the known actions of L and $L\lambda_L$ (up to the aforementioned scalar ambiguity). For, as already noted at the end of **4.4.1**, the desired map $\lambda^\bullet|_Q$ must send $\langle b_1 \rangle$ to $\langle r(e_1) \rangle$; we may assume that $b_1\lambda^\bullet = r(e_1)$. Recall that, in **4.4.3**(1), we obtained $l'(i) \in L\lambda_L$, $l(i) = l'(i)\lambda_L^{-1} \in L$ such that, in view of Lemma 4.6(a), $r(e_i) = r(e_1)^{l'(i)}$ and $b_i = b_1^{l(i)}$ for each i. Then we must also have $b_i\lambda^\bullet = (b_1^{l(i)})\lambda^\bullet = r(e_1)^{l(i)\lambda^\bullet} = r(e_i)$ for each i.

Also, we know that $\lambda^\bullet$ must be semilinear, but in fact it is linear: use the elements s_0' and s_0 (cf. **4.4.2**) as in the argument within Lemma 3.14. Moreover, since we know that the forms on Q and "Q" must be preserved up to a scalar and a field automorphism, while $\lambda^\bullet$ is linear on Q, we can use $\lambda^\bullet$ to determine the desired form on Q.

Now (i) is clear, and (ii) follows from linearity as in the proof of Lemma 3.14. $\square$

Remark. The spaces Q and $V_L = \langle \alpha, \gamma \rangle^\perp$ are related via $b = \sum_i k_i b_i \mapsto \dot{b} = \sum_i k_i e_i$, so that
$$b\lambda^\bullet = r(\dot{b}) \quad \text{for all } b \in Q.$$

We have three orthogonal spaces: $V_L = \langle \alpha, \gamma \rangle^\perp$, Q and "Q". We have seen that $\lambda^\bullet \colon Q \to$ "Q" is an isometry; so is $r \colon V_L \to$ "Q" (cf. **4.1.3**); hence, so also is $b \mapsto \dot{b}$ in view of the above relation. The form on V_L was called φ' in **4.4.1**; let φ'' denote the one on "Q". Then $\varphi_Q(b) = \varphi'(\dot{b}) = \varphi''(b\lambda^\bullet)$ for all $b \in Q$.

4.5   The homomorphism λ and straight-line programs

Points were defined in **4.3.2**.

4.5.1   Labeling points

We now label any given point by a 1–space in $V := \mathbb{F}^d$ in a G_α'–invariant manner.

Label Q by the 1–space $\langle 0, \ldots, 1, 0 \rangle$ and $Q(\gamma) = Q^{j(\gamma)}$ by $\langle 0, \ldots, 0, 1 \rangle$.

Consider any point $Q(\delta)$ not perpendicular to Q (cf. Lemma 4.15). Use Lemma 4.17 to find the unique $b \in Q$ conjugating $Q(\gamma)$ to $Q(\delta)$, and find $b\lambda^\bullet$ using Lemma 4.20(i). Then label $Q(\delta)$ as $\langle 0, \ldots, 0, 1 \rangle (b\lambda^\bullet) = \langle 0, \ldots, 0, 1 \rangle r(\dot{b})$ $= \langle \dot{b}, -\varphi_Q(b), 1 \rangle = \langle \dot{b}, -\varphi'(\dot{b}), 1 \rangle$.

Next consider a point $Q(\beta) = Q^g \neq Q$ perpendicular to Q. Clearly Q is perpendicular to $Q^{g^{-1}}$; find $1 \neq u \in Q \cap Q^{g^{-1}}$ using Lemma 4.15. In **4.3.3** we found two generators of L; find one of these generators h such that $Q^{g^{-1}h}$ is not perpendicular to $Q^{g^{-1}}$ (although it is certainly perpendicular to Q; note that L moves $\alpha^{\perp} \cap (\alpha^{g^{-1}})^{\perp}$). If $Q^{g^{-1}hg}$ and $Q^{g^{-1}hug}$ are labeled $\langle z_1, -\varphi'(z_1), 1\rangle$ and $\langle z_2, -\varphi'(z_2), 1\rangle$, respectively, then label $Q(\beta)$ as $\langle z_1 - z_2, -\varphi'(z_1) + \varphi'(z_2), 0\rangle$. It is straightforward to check that this labeling is forced by that of the points not perpendicular to Q. (Since $\alpha^{g^{-1}h} \notin (\alpha^{g^{-1}})^{\perp}$ we have $\alpha^{g^{-1}hu} \neq \alpha^{g^{-1}h}$. Now $\beta = \alpha^g, \alpha^{g^{-1}hg}$ and $\alpha^{g^{-1}hug}$ are distinct and collinear, and the last two are not in $((\alpha^{g^{-1}})^g)^{\perp} = \alpha^{\perp}$.)

Time: $O(\mu q e d^2)$ to label one point (see **4.4.3**(4) for the disappearance of $\log d$ factors here).

Remark. Implicit in the above is the fact that there is an epimorphism from G onto a simple orthogonal group sending triples of collinear points to triples of collinear points.

4.5.2   The homomorphism and the form

We now construct a homomorphism λ from G onto $\mathrm{P}\Omega(V)$.

By **4.4.4**, the label of $Q(\gamma)^{b_i}$ is $\langle v_i \rangle$ with $v_i := (e_i, -1, 1)$, since $\varphi'(e_i) = 1$. (These vectors have the property that they will be singular once we have the quadratic form on V. This explains their slightly unpleasant appearance.)

Let $b_0 := \prod_1^{d-2} b_i$ and $b_* := b_j b_{j'}$, where $\varphi'(e_j + e_{j'}) \neq 2$ (j and j' exist in view of the construction in Lemma 4.19). Then the label of $Q(\gamma)^{b_0}$ is $\langle v_0 \rangle$, where $v_0 := (\sum_i e_i, -\varphi'(\sum_i e_i), 1)$, while that of $Q(\gamma)^{b_*}$ is $\langle v_* \rangle$, where $v_* := (e_j + e_{j'}, -\varphi'(e_j + e_{j'}), 1)$.

Now suppose that we are given $x \in G$; we wish to find a corresponding matrix $\hat{x}$. Find each image $Q(\gamma)^{b_i x}$, $0 \leq i \leq d$ or $i = *$, and use **4.5.1** to find its label $\langle w_i \rangle$. Let $v_i \hat{x} = k_i w_i$ for an unknown matrix $\hat{x}$ and unknown nonzero scalars k_i to be determined. Then

$$
\begin{aligned}
k_0 w_0 &= v_0 \hat{x} = (\sum_1^{d-2} e_i, -\varphi'(\sum_1^{d-2} e_i), 1)\hat{x} \\
&= \sum_1^{d-2} k_i w_i + [-\varphi'(\sum_1^{d-2} e_i) + d - 2]k_{d-1}w_{d-1} + (3-d)k_d w_d
\end{aligned}
$$

uniquely determines the scalars k_i/k_0, $i \leq d - 2$. Similarly, $k_* w_* = v_* \hat{x} = k_j w_j + k_{j'} w_{j'} + (2 - \varphi'(e_i + e_{j'}))k_{d-1}w_{d-1} - k_d w_d$ determines $\hat{x}$ up to a scalar matrix, and hence produces the desired homomorphism $\lambda: x \mapsto \hat{x}$ modulo scalars.

As in Section 3, this scalar ambiguity is essential when $G \cong \mathrm{P}\Omega(V)$, but when $G \cong \Omega(V)$ it will disappear once we have proved Proposition 4.23.

Consistency of λ and λ_L. As in **3.5.2**, modulo scalars we have

$$x\lambda = \begin{pmatrix} x\lambda_L & O \\ O & I \end{pmatrix} \tag{4.21}$$

for any $x \in L$. For, if $b \in Q$, then **4.5.1** and **4.4.4** together with $(Q(\gamma)^b)^x = Q(\gamma)^{b^x}$ yield $\langle \dot{b}, -\varphi'(\dot{b}), 1\rangle x\lambda = \langle \dot{b}^x, -\varphi'(\dot{b}^x), 1\rangle = \langle \dot{b}(x\lambda_L), -\varphi'(\dot{b}), 1\rangle$. Since $L\lambda = (L\lambda)'$ fixes $\langle v_{d-1}\rangle$ and $\langle v_d\rangle$, this proves our assertion.

We also note that, modulo scalars, λ *and* $\lambda^\bullet$ *coincide on* Q. For, let $b \in Q$. Then $Q^b = Q$, so that $b\lambda$ fixes $\langle v_{d-1}\rangle$; $\langle v_d\rangle(b\lambda) = \langle \dot{b}, -\varphi'(\dot{b}), 1\rangle$; and, for any $c \in Q$, from $(Q(\gamma)^c)^b = Q(\gamma)^{cb}$ we deduce that $\langle \dot{c}, -\varphi'(\dot{c}), 1\rangle(b\lambda) = \langle \dot{c} + \dot{b}, -\varphi'(\dot{c} + \dot{b}), 1\rangle$. On the other hand, $b\lambda^\bullet$ also fixes v_{d-1}, and (using (4.3))

$$
\begin{aligned}
(\dot{c}, -\varphi'(\dot{c}), 1)(b\lambda^\bullet) &= (\dot{c}, 0, 0)r(\dot{b}) - \varphi'(\dot{c})v_{d-1}r(\dot{b}) + v_d r(\dot{b}) \\
&= (\dot{c}, -(\dot{c}, \dot{b})', 0) - \varphi'(\dot{c})v_{d-1} + (\dot{b}, -\varphi'(\dot{b}), 1) \\
&= (\dot{c} + \dot{b}, -\varphi'(\dot{c} + \dot{b}), 1),
\end{aligned}
$$

as asserted.

Recursively finding $\mathcal{S}^*\lambda$: As in **3.5.2**, by (4.21) we can obtain $\mathcal{S}^*(1)\lambda = \mathcal{S}_L^*\lambda$ from $S_L^*\lambda_L$ by bordering; we just saw that the $O(ed)$ elements of $\mathcal{S}^*(2)\lambda$ can be found in $O(ed \cdot \mu qd)$ time using $\lambda^\bullet$; and $\mathcal{S}^*(3)\lambda$ can be found using e applications of **4.5.1**. **Time:** $O(ed \cdot \mu qed^2)$ *to find $\mathcal{S}^*\lambda$ recursively, in addition to the time to find S_L^*.*

Finding $\mathcal{S}^*\lambda$ in Theorem 1.1': As in **3.5.2**, raise matrices to the power $1 - q$ in order to find the exact image $x\lambda$ of each $x \in \mathcal{S}^*$. **Time:** $O(d^3 \cdot d^2 e \log q)$.

Finding a quadratic form: By (4.21), L acts on $(V_L, 0, 0)$ as it does on V_L, so that the quadratic form on $(V_L, 0, 0)$ arises from a scalar multiple of the form φ' on V_L; we may assume that this scalar is 1. In view of the labels for Q and $Q(\gamma)$ in **4.5.1**, since each point $\langle \dot{b}, -\varphi'(\dot{b}), 1\rangle$ is singular this determines the form φ on V.

Time: $O(\mu qd)$ to determine $\varphi(v)$ for any given $v \in V$: write $v = (u, k, l)$ with $u \in V_L$, and then $\varphi(v) = \varphi'(u) + kl$.

4.5.3 Theorems 1.1(iii,iv) and 1.1'(iii,iv)

Proposition 4.22 *In deterministic $O(d^4 \log q)$ time, given $h \in \mathrm{GL}(d, q)$,*

(a) *it can be determined whether or not h (modulo scalars) is in $G\lambda$, and*

(b) *if so then a straight-line program of length $O(d^2 \log q)$ can be found from $\mathcal{S}^*\lambda$ to h (only modulo scalars in the situation of* Theorem 1.1(iv))*, after which $h\lambda^{-1}$ can be found in additional $O(\mu d^2 \log q)$ time.*

Moreover,

(c) *in* Theorem 1.1′, $Z(G)$ *can be found in $O(\mu d^2 \log q)$ time.*

Proof. We reorder our basis $v_1, \ldots, v_d$ as $e_1, \ldots, e_d$ with $e_1 = v_{d-1}$ and $e_d = v_d$. Then $e_2, \ldots, e_{d-1} \in \langle e_1, e_d \rangle^\perp$ and $L\lambda$ is 1 on $\langle e_1, e_d \rangle$. We will write matrices with respect to this basis.

Recall that $[V, Q\lambda] = \langle e_1 \rangle$ and $[V, Q(\gamma)\lambda] = \langle e_d \rangle$, where $Q(\gamma) = Q^{j(\gamma)}$. The group $Q\lambda$ is given in (4.3). We have a GF(p)–basis $\mathcal{B}_p\lambda$ of this group, and this lets us express its elements using straight-line programs from $\mathcal{S}^*\lambda$ in $O(d^3 \log q)$ time. A similar statement holds for $Q(\gamma)\lambda$.

(a) Let $h \in \mathrm{GL}(d, q)$. For membership testing use an $O(d^3)$ algorithm for the determinant and another one for the spinor norm employing Wall forms of the generators of G [Ta, p. 163].

(b) If $-h \in \Omega(V)$ then replace h by $-h$.

If $[Q(\gamma)\lambda]^h$ is not perpendicular to $Q\lambda$ (tested using the form on V found at the end of **4.5.2**), write a matrix $u \in Q\lambda$ with $[Q(\gamma)\lambda]^{hu} = Q(\gamma)\lambda$ (cf. (4.3)), and find a straight-line program of length $O(d \log q)$ from $\mathcal{S}^*(2)\lambda$ to u. Thus, in this case we may replace h by hu.

On the other hand, if $[Q(\gamma)\lambda]^h$ is perpendicular to $Q\lambda$ then (as in (4.3)) write a matrix $u \in Q(\gamma)\lambda$ such that $[Q(\gamma)\lambda]^{hu}$ is not perpendicular to $Q\lambda$, and proceed as before.

Thus, we may assume that h normalizes $Q(\gamma)\lambda = [Q\lambda]^{j(\gamma)\lambda}$. Now replace h by $[j(\gamma)\lambda]h[j(\gamma)\lambda]^{-1}$ in order to have $[Q\lambda]^h = Q\lambda$ (cf. **4.2.2**(iv)).

We can write $h = \begin{pmatrix} a^{-1} & O & 0 \\ * & h' & O \\ * & * & a \end{pmatrix}$ for some $a \in \mathbb{F}^*$ and a $d - 2 \times d - 2$ matrix h' representing an isometry of $\langle e_1, e_d \rangle^\perp$. Similarly, each element $j \in J\lambda$ normalizing Q also can be written $j := \begin{pmatrix} a' & O & 0 \\ * & j' & O \\ * & * & a'^{-1} \end{pmatrix}$; use **2.3** to find one such that $|a'| = q - 1$ and then to find a power of this one for which $a' = a$.

Then $hj = \begin{pmatrix} 1 & O & 0 \\ * & h'j' & O \\ * & * & 1 \end{pmatrix}$, where $h'j'$ represents an element of $\Omega^\varepsilon(d - 2, q)$.

Recursively find $b := \begin{pmatrix} 1 & O & 0 \\ O & (h'j')^{-1} & O \\ 0 & O & 1 \end{pmatrix}$ using a straight-line program from $\mathcal{S}_L^*\lambda$. Then $hjb \in Q\lambda$, and hence hjb can be found as above using a straight-line program from $\mathcal{S}^*(2)\lambda$ of the desired length.

(c) As in Proposition 3.17, $Z(G) = Z(G\lambda)\lambda^{-1}$. Hence, apply (b) to find a straight-line program to $Z(G\lambda)$ and then also one to $Z(G)$. (N.B.—$Z(G\lambda) \neq 1$ if and only if $Z(G\lambda) = \langle -1 \rangle$, q is odd, d is even, and one of the following holds: $d \equiv 0 \pmod 4$ and $G\lambda = \Omega^+(d,q)$, or $d \equiv 2 \pmod 4$, $q \equiv -\varepsilon \pmod 4$, and $G\lambda = \Omega^\varepsilon(d,q)$.) $\square$

Proposition 4.23 *In deterministic $O(\mu q e d^2)$ time, a straight-line program of length $O(d^2 \log q)$ can be found from $\mathcal{S}^*$ to any given $g \in G$. (Hence, the corresponding straight-line program can be written from $\mathcal{S}^*\lambda$ to $g\lambda$, and $g\lambda$ can be computed in additional $O(d^3 \cdot d^2 \log q)$ time.)*

Proof. Once again we begin by modifying g so that it normalizes Q.

Define $y \in \{1, j(\gamma)\} \cup \mathcal{S}^*(2)^{j(\gamma)}$ such that Q^{gy} and Q are not perpendicular, as follows: if Q^g and Q are not perpendicular let $y = 1$; if Q^g and $Q^{j(\gamma)}$ are not perpendicular let $y = j(\gamma)^{-1}$; if Q^g is perpendicular to both Q and $Q^{j(\gamma)}$, test the members of $\mathcal{S}^*(2)^{j(\gamma)} \subset Q(\gamma)$ (considering at most one per root group) in order to find one of them, y, not normalizing Q^g (so that y moves α^g to a point of $\langle \alpha^g, \gamma \rangle$ not in $\alpha^\perp$). Then we have a short straight-line program from $\mathcal{S}^*$ to y (cf. **4.2.2**(iv)). Use Lemma 4.17 to find $u \in Q$ such that $Q^{gyu} = Q^{j(\gamma)}$, and use **4.4.3**(2) to find a straight-line program of length $O(d \log q)$ from $\mathcal{S}^*(2)$ to u. Then $gyuj(\gamma)^{-1}$ normalizes Q, and if we can find a straight-line program of length $O(d^2 \log q)$ from $\mathcal{S}^*$ to this element then we can find one to g as well (cf. **4.2.2**(iv)).

Similarly, use Lemma 4.17 to find $v \in Q$ such that $Q(\gamma)^{gv} = Q(\gamma)$, use **4.4.3**(2) to obtain v from $\mathcal{S}^*(2)$ using a short straight-line program, and replace g by gv. Now we have reduced to the case in which g normalizes both Q and $Q(\gamma)$.

Use **4.4.3**(5) to find the matrix $\tilde{g}$ representing the linear tranformation of Q induced by the action of g. Recall that we have the quadratic form φ_Q on Q (cf. Lemma 4.20(i)). Let $v \in Q$ be a nonsingular element of Q (e.g., b_1 or $b_1 b_{-1}$ for q odd or even, respectively). Find $\varphi(v)$ and $\varphi(v^{\tilde{g}})$; by Lemma 4.6(a) we have $\varphi(v^{\tilde{g}}) = l^2 \varphi(v)$ for some $l \in \mathbb{F}$ depending only on g. Use the exponents in our Zech table in **2.3** to find a power $(s^+)^i$ of the element in **4.2.2**(vi) such that $\varphi(v^{\tilde{g}(\tilde{s}^+)^i}) = \varphi(v)$ (here $\tilde{s}^+$ is found using **4.4.3**(5) once again). Then $\tilde{g}(\tilde{s}^+)^i$ is an isometry of Q, we can find a straight-line program from $\mathcal{S}^*$ to $(s^+)^i$ of length $O(\log q)$ (cf. **4.2.2**(vi)) and hence we can replace g by $g(s^+)^i$.

Now $\tilde{g}$ is an isometry of Q; we wish to arrange to have it inside $\Omega(Q)$. Test for this property using both a determinant and a spinor norm computation (again using Wall forms [Ta, p. 163]); each takes time $O(d^3)$. By Lemma 4.6(b), if $\tilde{g} \notin \Omega(Q)$ then $q \equiv 3 \pmod 4$ and, if $h := (s^+)^{(q-1)/2}$, we also have $\tilde{h} \notin \Omega(Q)$ and a straight-line program from $\mathcal{S}^*$ to h of length $O(\log q)$ (cf. **4.2.2**(vi)). Hence we can replace g by gh in order to have $\tilde{g} \in \Omega(Q)$.

Use Proposition 4.22(b) to find $l := \tilde{g}\lambda_L^{-1}$ by a straight-line program of length $O(d^2 \log q)$ from $\mathcal{S}_L^*$. By Lemma 4.20(ii), $\tilde{l} = l\lambda_L = \tilde{g}$, so that $c := gl^{-1}$ induces 1 on Q/T, normalizes Q and $Q(\gamma)$, and hence lies in $Z(G)$. Thus, we have obtained an element $l \in gZ(G)$ from $\mathcal{S}^*$. The algorithm ends here in the case of Theorem 1.1. For Theorem 1.1′, use Proposition 4.22(c) to find a short straight-line program from $\mathcal{S}^*$ to c. Thus, we have obtained g by a straight-line program of length $O(d^2 \log q)$ from $\mathcal{S}^*$, as required.

Time: $O(\mu q e d^2)$, dominated by Lemma 4.17 and **4.4.3**(5). $\square$

4.6   Small dimensions and total time

In this section we will complete the proof of Theorems 1.1 and 1.1′ for the orthogonal groups $P\Omega^\varepsilon(d, q)$ and $\Omega^\varepsilon(d, q)$, *assuming* that we know d and q (except for the additional verification using a presentation; cf. **7.2.2**).

4.6.1   From $\mathrm{PSL}(4, q)$ to $P\Omega^+(6, q)$

It is straightforward to pass from the projective 4–dimensional representation of $\mathrm{PSL}(4, q)$ on a vector space V to the 6–dimensional one on $V \wedge V$ as $P\Omega^+(6, q)$. Passing from a linear transformation of V to one of $V \wedge V$ takes $O(1)$ time.

4.6.2 $d = 7$ or 8

This is a refinement that can be handled reasonably quickly by brute force (cf. **3.2.3**). *We will handle these cases in $O(\xi q e + \mu q \log^2 q)$ time.*

We may assume that $q \geq 49$. In particular, $|\Omega^+(2, q)| = (q-1)/(2, q-1) > 4$. Choose up to 320 elements to find $\tau \in G$ with $|\tau| = w_0 y_0 z_0$ as follows:

G	w_0	z_0	y_0	V
$P\Omega(7,q)$	$\mathrm{ppd}^{\#}(p;e)$	$\mathrm{ppd}^{\#}(p;4e)$	1	$V_2^+ \perp V_4^- \perp V_1$
$P\Omega^-(8,q)$	$\mathrm{ppd}^{\#}(p;e)$	$\mathrm{ppd}^{\#}(p;6e)$	1	$V_2^+ \perp V_6^-$
$P\Omega^+(8,q)$	$\mathrm{ppd}^{\#}(p;e)$	$\mathrm{ppd}^{\#}(p;2e)$	$\mathrm{ppd}^{\#}(p;4e)$	$V_2^+ \perp V_2^- \perp V_4^-$

We make the additional restriction that $8 \big| |\tau|$ if q is a Mersenne or Fermat prime. For $P\Omega^+(8,q)$ there is more than one way τ can act, but all ways are equivalent under $\mathrm{Aut}\,P\Omega^+(8,q)$ and lead to modules that are permuted transitively by this automorphism group; therefore, we may assume that the action on V in the last row is as stated up to triality. There is no such ambiguity in the action of τ for the remaining cases. In all cases it follows from the decomposition of V that we obtain such an element with probability $> 1 - 1/4d^2$ (the probability is $\geq 1/2^6$ for one choice in view of **4.1.5**).

Let $w := q - 1$. When $z_0 = \mathrm{ppd}^{\#}(p;2ei)$ for $i = 1, 2$ or 3 write $z := q^i + 1$, and proceed similarly for y. Then $|\tau| \big| wyz$, and the integers w, z, y pairwise have g.c.d ≤ 2. Then $a := \tau^{\mathrm{lcm}(y,z)}$ induces an element of $\mathrm{ppd}^{\#}(p;e)$–order on V_2^+ and 1 on $V_2^{+\perp}$.

Now select up to $2^{14}10$ pairs a_1, a_2 of conjugates of a. Then $J := \langle a, a_1, a_2 \rangle$ is $\Omega^+(6,q)$ with probability $> 1/2^{10}10$ for each choice, by Lemma 4.12(ii). If indeed $J \cong \Omega^+(6,q)$ then the algorithm in Section 3 combined with **4.6.1** and **7.2.2** recognizes this fact with probability $> 3/4$. Hence, with probability $> 1 - 1/8^3 \geq 1 - 1/8d^2$, for at least one of our choices a_1, a_2 we obtain a 6–dimensional space V_J, a generating set $\mathcal{S}_J^*$ consisting of root elements, and an isomorphism $\lambda_J \colon J \to \Omega(V_J) = \Omega^+(6,q)$. Use λ_J^{-1} to find the stabilizer J_α of a singular point α chosen in the 2–dimensional support V_2^+ of a within the 6–space V_J. Then τ fixes α.

Use λ_J^{-1} to find the subgroup G_6 generated by the long root elements of J_α and to find $Q_6 := O_p(G_6)$. Then $G_6/Q_6 \cong \Omega^+(4,q)$.

Let $G'_\alpha := \langle G_6^{\tau^i} \mid 0 \leq i \leq d \rangle$ and $Q := \langle Q_6^{\tau^i} \mid 0 \leq i \leq d \rangle$. The argument in Lemma 4.14 shows that $Q = \text{'}Q\text{'}$ and $G'_\alpha = \mathrm{N}_G(Q)'$.

Now proceed as before, starting from **4.3.2**.

Reliability: $> 1 - 1/2d^2$.

Time: $O(\xi qe + \mu q \log^2 q)$, with most of the time spent either using Section 3 for J or in **4.5.2**. Note that $O(\mu qe)$ time is spent on **4.4–4.5**.

4.6.3 $P\Omega^-(6,q)$

We have not encountered the group $P\Omega^-(6,q)$ in this section. Since it is isomorphic to $PSU(4,q)$ it will be handled later in **6.6.2**. However, in that section the timing involves q^2 in order to obtain the unitary space. Here we wish to provide a faster algorithm in order to obtain the 6–dimensional orthogonal space. We may assume that $q \geq 17$.

Choose up to $2^5 5$ elements τ in order to find one of $\mathrm{ppd}^\#(p;e)\mathrm{ppd}^\#(p;4e)$–order, so that $a := \tau^z$ has $\mathrm{ppd}^\#(p;e)$–order where $z := q^2 + 1$, and τ^{q-1} has $\mathrm{ppd}^\#(p;4e)$–order.

Choose up to $2^6 640$ conjugates a' of a, and for each use **3.6.2** to test whether $J := \langle a, a' \rangle \cong \Omega^+(4,q) \cong SL(2,q)\circ SL(2,q)$, in which event find a generating set $\mathcal{S}_J^*$ consisting of long root elements, an isomorphism $\lambda_J\colon J \to \Omega^+(4,q) = \Omega(V_J)$, and the following:

(i) The stabilizer J_α of a singular point α of $[V_J, a\lambda_J]$;

(ii) $Q_4 := O_p(J_\alpha)$, of order q^2, with a generating set consisting of nontrivial long root elements; let r be one of these elements;

(iii) $b \in N_J(Q_4)_\alpha$ of order $q-1$ such that $[V_J, a\lambda_J] \neq [V_J, b\lambda_J]$; and

(iv) an element $j(\gamma)$ such that $j(\gamma)\lambda_L$ sends α to the singular point $\gamma \neq \alpha$ of $[V_J, b\lambda_J]$, together with a straight-line program of length $O(1)$ from $\mathcal{S}_J^*$ to $j(\gamma)$.

Define $Q := \langle Q_4^{\tau^{(q-1)i}} \mid 0 \leq i \leq 3 \rangle$ and $G_\alpha' := \langle Q, \tau^{2(q-1)}, \tau^{2(q-1)b} \rangle$. Then $Q = {}'Q'$ (by Lemma 2.7) and $G_\alpha'/Q \cong \Omega^-(4,q) \cong PSL(2,q^2)$ (since b was chosen so that $\langle \tau^{2(q-1)} \rangle^b \neq \langle \tau^{2(q-1)} \rangle$).

Define $Q(\gamma) := Q^{j(\gamma)}$.

Lemmas 4.15–4.17 hold as before.

Corollary 4.18: As in that corollary, use Lemma 4.17 to find $u \in Q$ such that $\tau^{2(q-1)bu}$ normalizes $Q(\gamma)$, and obtain $L := \langle \tau^{2(q-1)}, \tau^{2(q-1)bu} \rangle < G_\alpha'$ normalizing $Q(\gamma)$ and such that $G_\alpha' = Q \rtimes L$. Then $L \cong \Omega^-(4,q)$. We wish to recognize L constructively; since using **3.6.1** would involve q^2 in the timing, we proceed indirectly as follows.

Choose up to $2^2 5$ elements $l \in L$ in order to find one of $\mathrm{ppd}^\#(p;e)$–order. Choose up to $2^6 640$ conjugates a^x of a in G, and for each use **3.6.2** to try to find an isomorphism $\langle l^{q+1}, a^x \rangle \cong \Omega^+(4,q)$; when this is obtained use the isomorphism to find $y \in \langle l^{q+1}, a^x \rangle$ such that $\langle a^x \rangle^y = \langle a \rangle$. Then $a \in L^g$, where

$g := (xy)^{-1}$, and $M := \langle L^g, r \rangle \cong \Omega(5, q)$ for r in (ii). Use up to 2 applications of **5.6.1** and **5.6.3** to find an isomorphism $\lambda_M \colon M \to \Omega(5, q)$; use it to find an isomorphism $L^g \to \Omega^-(4, q)$ and then another one $\lambda_L \colon L \to \Omega^-(4, q) = \Omega(V_L)$.

Use λ_L to replace the original generating set for L by a generating set $\mathcal{S}_L^*$ consisting of $O(\log q)$ root elements (these are not long root elements).

Now proceed as before, starting with **4.3.3**.

Correctness: We used the power $\tau^{2(q-1)}$ in order to guarantee that L lies in $(G_\alpha)'$; and two cyclic subgroups of $\mathrm{PSL}(2, q^2)$ of $\mathrm{ppd}^\#(p; 4e)$–order generate $\mathrm{PSL}(2, q^2)$. Since $l \in \Omega^-(4, q)$ we have $|l^{q+1}| \, | \, q - 1$, so the elements l^{q+1} and a^x behave as required in Lemma 4.12(i). Note that $[V, r] \cap [V, L^g] = [V, r] \cap [V, a]$ can be viewed as the singular point α, and L^g is transitive on the totally singular lines meeting $[V, L^g]$ in that point. Then $\langle L^g, r \rangle$ lies in some subgroup $\Omega(5, q)$, and hence is such a subgroup since L^g already is $\Omega^-(4, q)$.

Reliability: $> 1 - 5/2^5$: a single choice of τ has the desired order with probability $\geq 1/2^5$ in view of **4.1.5**; a single choice of a' has satisfies $\langle a, a' \rangle \cong \Omega^+(4, q)$ with probability $\geq 1/640$ by Lemma 4.12(i), and then **3.6.2** produces an isomorphism with probability $\geq 1/2$; similar statements hold for a single choice of a^x; a single choice of l has the desired order with probability $\geq 1/4$; and each application of **5.6.1** produces an isomorphism with probability $\geq 1 - 1/2 \cdot 2^2$.

Time: $O(\xi q e + \mu q \log^2 q)$, dominated by **3.6.2** and **5.6.1**.

4.6.4 Total time and reliability

Part (i) of Theorems 1.1 and 1.1′ is in Section 7; (ii) is **4.5.2**; (iii) and (iv) are in **4.5.3**; and (vi) is in **4.3.3**. For (vii), by tallying and recalling the recursive call we find that the *total time is*

$$O(d\{\xi q[d + e]\log d + \mu q d^3 \log^2 q \log^3 d\}),$$

dominated by **4.2.1** and **4.2.2** (recall that factors $\log d$ in timings in all but the last corollary in **4.3.2** were removed in **4.4.3**(4)). By **4.5.3**, the straight-line algorithms for (iii) and (iv) take time $O(\mu q e d^2 + d^5 \log q)$ and $O(\mu d^2 \log q)$, respectively.

Reliability: $> 1/2$.

5 Symplectic groups: $\mathrm{PSp}(2m, q)$

In this section we will prove Theorems 1.1 and 1.1′ when G is $\mathrm{PSp}(2m, q)$ or $\mathrm{Sp}(2m, q)$, $2m \geq 4$, arising from the group of isometries of a nonsingular alternating bilinear form $(\ ,\ )$ on a $2m$–dimensional vector space V over $\mathrm{GF}(q)$. When the characteristic p is odd, this is probably the easiest case of Theorem 1.1′. However, when $p = 2$ there are complications caused by the fact that G behaves like the isomorphic orthogonal group $\Omega(2m + 1, q)$.

In the characteristic 2 case we let V' denote a $2m + 1$–dimensional orthogonal module for G, with radical $\mathrm{rad}V'$, such that $V'/\mathrm{rad}V' \cong V$ as G–modules. In some places we will switch our focus to V' and simply use algorithms from the preceding section.

If α is any point then $\alpha^{\perp}/\alpha$ is a symplectic space of dimension $2m - 2$.

5.1 Properties of G

5.1.1 Transvections and root groups

For each point $\alpha = \langle u \rangle$, G_{α} has a normal subgroup consisting of transvections, the *transvection group*

$$T(\alpha) = \{v \mapsto v + k(v, u)u \mid k \in \mathrm{GF}(q)\} \cong \mathrm{GF}(q)^{+}. \qquad (5.1)$$

While $T(\alpha)$ is a "long root group", we will also need the *short root group* determined by any distinct perpendicular points $\langle u_1 \rangle$, $\langle u_2 \rangle$:

$$R(\langle u_1 \rangle, \langle u_2 \rangle) = \{v \mapsto v + k(v, u_1)u_2 + k(v, u_2)u_1 \mid k \in \mathrm{GF}(q)\} \cong \mathrm{GF}(q)^{+}. \qquad (5.2)$$

Note that $R(\langle u_1 \rangle, \langle u_2 \rangle)$ induces 1 on $\langle u_i \rangle^{\perp}/\langle u_i \rangle$ for $i = 1, 2$. This subgroup depends on the choice of points $\langle u_1 \rangle$, $\langle u_2 \rangle$ of $\langle u_1, u_2 \rangle$ when p is odd but not when $p = 2$. Moreover, each nontrivial $r \in R(\langle u_1 \rangle, \langle u_2 \rangle)$ *completely determines* $R(\langle u_1 \rangle, \langle u_2 \rangle)$. Namely, when $p = 2$ any two points of $[V, r] = \langle u_1, u_2 \rangle$ determine the same group $R(\langle u_1 \rangle, \langle u_2 \rangle)$; when p is odd the assertion follows from the fact that $\langle u_1 \rangle$ and $\langle u_2 \rangle$ are the only points β such that r induces 1 on $\beta^{\perp}/\beta$.

5.1.2 Commutator relations

If α and β are points, then

$[T(\alpha), T(\beta)] = 1$ if α and β are perpendicular, in which case $T(\alpha) \cup T(\beta)$ is the set of transvections in $\langle T(\alpha), T(\beta) \rangle$, and

92

$\langle T(\alpha), T(\beta) \rangle$ is $\mathrm{Sp}(2, q) \cong \mathrm{SL}(2, q)$ if α and β are not perpendicular. In this case $V = \langle \alpha, \beta \rangle \perp \langle \alpha, \beta \rangle^{\perp}$, and $\langle T(\alpha), T(\beta) \rangle$ induces $\mathrm{Sp}(2, q)$ on the first summand and 1 on the second one.

5.1.3 Q

For any point α let $Q = Q(\alpha)$ denote the normal subgroup of G_{α} consisting of all isometries inducing 1 on both α and $\alpha^{\perp}/\alpha$. If $\alpha = \langle u \rangle$ then Q consists of all maps

$$v \mapsto v + (z + ku, v)u + (u, v)z \quad \text{with} \quad z \in \alpha^{\perp}, \ k \in \mathrm{GF}(q).$$

Those with $z = 0$ comprise $T = T(\alpha)$, and $Q/T \cong \alpha^{\perp}/\alpha$. Note that all maps mentioned in **5.1.1** are conjugate to elements of $Q(\alpha)$.

Alternatively, if $(u, w) = 1$ then Q consists of all maps

$$r(z, l) : u \mapsto u, \quad v \mapsto v - (v, z)u \ \forall v \in \langle u, w \rangle^{\perp}, \quad w \mapsto w + z + lu \qquad (5.3)$$

for $z \in \langle u, w \rangle^{\perp}$ and $l \in \mathrm{GF}(q)$. Once again Q consists of *sparse matrices* with respect to a basis consisting of u, w and a basis of $\langle u, w \rangle^{\perp}$. We have

$$\begin{aligned} r(z, l)r(z', l') &= r(z + z', l + l' - (z, z')) \quad \text{and} \\ [r(z', l'), r(z, l)] &= r(0, 2(z, z')) \end{aligned} \qquad (5.4)$$

for all $z, z' \in \langle u, w \rangle^{\perp}$ and $l, l' \in \mathrm{GF}(q)$. It follows that $\mathrm{Z}(Q) = T$ if q is odd, and Q is elementary abelian if q is even.

Moreover, Q *is regular on the set of all points not in* $\alpha^{\perp}$, and

$$G_{\alpha} = Q \rtimes G_{\alpha\gamma} \quad \text{for} \quad \gamma = \langle w \rangle. \qquad (5.5)$$

If $\rho \in GF(q)^{*}$ is a generator, define s by

$$\begin{aligned} s : u &\mapsto \rho u, \quad w \mapsto \rho^{-1}w, \quad v \mapsto v \ \forall v \in \langle u, w \rangle^{\perp}, \quad \text{so} \\ r(z, l)^{s} &= r(\rho z, \rho^{2}l) \quad \text{for all } z \text{ and } l. \end{aligned} \qquad (5.6)$$

This turns Q/T into a $\mathrm{GF}(q)$–space. We exclude the cases $2m = 4$, $q \leq 3$. Then $(G_{\alpha\gamma})'$ is the symplectic group induced by G_{α}' on $\alpha^{\perp}/\alpha$, and *the* $(G_{\alpha\gamma})'$*–modules* $\alpha^{\perp}/\alpha$, $\langle \alpha, \gamma \rangle^{\perp}$ *and* Q/T *are isomorphic*. In fact,

$$\text{If } g \in (G_{\alpha\gamma})' \text{ then } r(z, l)^{g} = r(z^{g}, l) \text{ for all } z \text{ and } l. \qquad (5.7)$$

Morever, G'_α is irreducible on Q/T. The alternating form inherited on the GF(q)–space Q/T is given by $(r(z,l)T, r(z',l')T) = (z,z')$; once again there is "geometry" within Q/T.

Remark. If $p = 2$ then Q is the natural module for $\Omega(2m - 1, q) \cong \mathrm{Sp}(2m - 2, q)$, with radical T; the action of G'_α is indecomposable if $2m \geq 4$ but $G \neq \mathrm{Sp}(4,2)$. In the characteristic 2 case, Q is probably best viewed using the description in **4.1.3**.

5.1.4 Probabilistic generation

Cases (ii-iv) of Theorem 3.6 can occur. However:

Lemma 5.8 *If q is odd then, with probability $> (1 - 1/q)(1 - 1/q^2)(1 - 2/q)$ $(1 - 1/q^2) > 1/10$, four nontrivial transvections generate a subgroup $J \cong \mathrm{Sp}(4,q)$ inducing $\mathrm{Sp}(4,q)$ on the nonsingular 4–space $[V, J]$ and 1 on $[V, J]^\perp$.*

Proof. Consider randomly and independently chosen transvections t_i with centers α_i, $i = 1, 2, 3, 4$. We may fix t_1. Then the conditions $\alpha_2 \notin \alpha_1^\perp$, $\alpha_3 \notin \langle \alpha_1, \alpha_2 \rangle$ and $\alpha_4 \notin (\mathrm{rad}\langle \alpha_1, \alpha_2, \alpha_3 \rangle)^\perp \cup \langle \alpha_1, \alpha_2 \rangle^\perp$ occur with probability $\geq 1 - 1/q$, $1 - 1/q^2$ and $1 - 2/q$, respectively.

Now $J = \langle t_1, t_2, t_3, t_4 \rangle$ acts irreducibly on $\langle \alpha_1, \alpha_2, \alpha_3, \alpha_4 \rangle$. By [Ka2], the only possibilities for J are $\mathrm{Sp}(4, p^i)$ with GF$(p^i) \subseteq$ GF(q) (recall that q is odd). As in the proof of Lemma 3.7, the probability that $J \neq \mathrm{Sp}(4,q)$ is at most

$$\sum_{e/i \text{ is prime}} |\mathrm{Sp}(4,q) : \mathrm{Sp}(4,p^i)| \frac{(p^{4i} - 1)^4}{(q^4 - 1)^4} < 1/q^2,$$

where $p^{4i} - 1$ is the number of nontrivial transvections in $\mathrm{Sp}(4, p^i)$. $\square$

Remark. This lemma in false for $p = 2$, since $\mathrm{O}^\pm(4, q)$ can be generated by four transvections but $\mathrm{Sp}(4,q)$ never can be. In this case, the same proof shows that five transvections probably generate a subgroup $J \cong \mathrm{Sp}(4,q)$, but there are two possible conjugacy classes of such subgroups, for one of which $\dim[V, J] = 5$. These remarks are most easily seen by using the $2m + 1$–dimensional module V'.

Lemma 5.9 *If q is even and $M \neq G$ is an irreducible subgroup of G generated by M–conjugate short root groups, then M either is a subgroup of a naturally embedded subgroup $\Omega^\pm(2m, q)$ of G, or is $G_2(q)$ embedded naturally in $\Omega(7, q) \cong \mathrm{Sp}(6, q)$.*

Proof. This follows from [Ka2] if G is viewed as $\Omega(2m+1,q)$, in which case short root groups become long root groups. □

Lemma 5.10   (i) *If $m \geq 3$ and if $q \geq 8$ is even then, with probability $\geq 1/640$, for two elements g_1, g_2 of the same $\mathrm{ppd}^{\#}(p;e)-$ or $\mathrm{ppd}^{\#}(p;2e)-$ order such that each $[V,g_j]$ is a nonsingular 2–space, $\langle g_1, g_2\rangle$ induces 1 on $[V, \langle g_1,g_2\rangle]^{\perp}$ and $\Omega^{+}(4,q)$ on the nonsingular 4–space $[V, \langle g_1,g_2\rangle]$.*

(ii) *If $q > 2$ is even and $2m \geq 6$ then, with probability $> 1/2^{10}10$, for three elements g_1, g_2, g_3 having the same $\mathrm{ppd}^{\#}(2;e)-$ or $\mathrm{ppd}^{\#}(2;2e)-$ order and such that each $[V,g_j]$ is a nonsingular 2–space, $\langle g_1, g_2, g_3\rangle$ induces $\Omega^{+}(6,q)$ on the nonsingular 6–space $[V, \langle g_1, g_2, g_3\rangle]$ and 1 on $[V, \langle g_1, g_2, g_3\rangle]^{\perp}$.*

(iii) *If $q = 2$ and $2m \geq 8$ then, with probability $> 1/8$, for two elements g_1, g_2 of order 5 such that each $[V,g_j]$ is a nonsingular 4–space, $\langle g_1, g_2\rangle$ induces $\Omega^{-}(8,2)$ on the nonsingular 8–space $[V, \langle g_1,g_2\rangle]$ and 1 on $[V, \langle g_1,g_2\rangle]^{\perp}$.*

(iv) *If $q = 2$ or 4 and $2m \geq 6$ then, with probability $> 1/q^2$, for two elements g_1, g_2 of order 3 such that each $[V,g_j]$ is a nonsingular 2–space, $\langle g_1, g_2\rangle$ induces $\Omega^{-}(4,q)$ on the nonsingular 4–space $[V, \langle g_1,g_2\rangle]$ and 1 on $[V, \langle g_1,g_2\rangle]^{\perp}$; and hence, in particular, $\langle g_1\rangle$ and $\langle g_2\rangle$ are conjugate in $\langle g_1, g_2\rangle$.*

(v) *If $q \geq 5$ is odd then, with probability $\geq 1/32$, for two elements g_1, g_2 of the same $\mathrm{ppd}^{\#}(p;e)-$order not dividing 4 (or $\mathrm{ppd}^{\#}(p;e)\cdot\mathrm{ppd}^{\#}(p;e/2)-$order if e is even, or order 8 if $q = 9$) such that each $[V,g_j]$ is a nonsingular 2–space, $\langle g_1, g_2\rangle$ induces 1 on $[V, \langle g_1,g_2\rangle]^{\perp}$ and $\mathrm{Sp}(4,q)$ on the nonsingular 4–space $[V, \langle g_1,g_2\rangle]$.*

Proof. If q is even we work in the $2n+1$–space V' and let $U_j = [V',g_j]$.

(i,ii,iii) Since $\dim U_j = 2$ or 4, the argument in Lemma 4.12 can be repeated, and yields the same result.

(iv) If $q = 2$ then the number of V_4^{-} containing U_1 is the number of hyperbolic 2–spaces in $U_1^{\perp}$, which is $> 2^{4m-6}/2$. In each such V_4^{-} there are 3 subspaces U_2 of the correct isometry type that intersect U_1 trivially; for each of these the corresponding group $\langle g_1, g_2\rangle$ is $\Omega^{-}(4,2) \cong A_5$. Hence, the desired probability is $> (2^{4m-6}/2)3/\{(2^{2m}-1)2^{2m-1}/(2+1)2\} > 1/4$.

A similar argument handles the case $q = 4$, where this time there are two possibilities for the irreducible group $\langle g_1, g_2\rangle$.

(v) The argument in Lemma 4.12(i) shows that $\langle g_1, g_2 \rangle$ is irreducible on the 4–space $[V, \langle g_1, g_2 \rangle]$ with probability $\geq 1/32$. When q is odd all irreducible subgroups of $\mathrm{Sp}(4, q)$ have long been known [Mi2]. None of them other than $\mathrm{Sp}(4, q)$ is generated by two elements g_1, g_2 of the stated order and for which each $[V, g_j]$ is a nonsingular 2–space. Thus, $\langle g_1, g_2 \rangle$ must be the desired group $\mathrm{Sp}(4, q)$. $\square$

5.1.5 Irreducible elements

$\mathrm{Sp}(2m, q)$ contains elements of order $q^m + 1$, and each such element is irreducible. As in Lemma 2.5, *at least* $|\mathrm{Sp}(2m, q)|/4m$ *elements of* $\mathrm{Sp}(2m, q)$ *have* $\mathrm{ppd}^{\#}(p; 2em)$–*order*.

5.2 Transvections and root elements

We now begin the symplectic case of Theorems 1.1 and 1.1′. We postpone the case $2m = 4$ until **5.6.1**. Thus, we *assume that* $2m \geq 6$. We also assume that $2m \geq 12$ if $q = 2$ and $2m \geq 8$ if $q = 4$.

5.2.1 Finding transvection groups and J: $p \neq 2$

This time we will present two approaches. The first is slightly faster and resembles what we will do for even q, but it does not work if q is too small; the second is simpler and works for all odd q.

Case $q > 5$: Choose up to $\lceil 32m \log(4m) \rceil$ elements $\tau \in G$ in order to find one whose order behaves as follows:

$$\begin{cases} \mathrm{ppd}^{\#}(p; e) \cdot \mathrm{ppd}^{\#}(p; 2e(m-1)) & \text{if } e \text{ is odd,} \\ \mathrm{ppd}^{\#}(p; e/2) \cdot \mathrm{ppd}^{\#}(p; e) \cdot \mathrm{ppd}^{\#}(p; 2e(m-1)) & \text{if } e \text{ is even but } q \neq 9, \\ 8\mathrm{ppd}^{\#}(3; 4(m-1)) & \text{if } q = 9; \end{cases}$$

if q is prime or the square of a Mersenne prime we also require that $8 \| |\tau|$. Then τ must split V as $V = V_2 \perp V_{2m-2}$, inducing elements of order dividing $w := q - 1$ on V_2 and $z := q^{m-1} + 1$ on V_{2m-2} since $(w, z) = 2$. As in **5.1.5**, a single choice of τ has the required order with probability $> (1/4)(1/4m)$; hence at least one of our choices has the desired order with probability $> 1 - 1/16m^2$, in which case $a := \tau^z$ has $\mathrm{ppd}^{\#}(p; e)$–order ≥ 3.

Choose up to $\lceil 2^7 \log(4m) \rceil$ conjugates b of a. For each conjugate let $J := \langle a, b \rangle$, and use Corollary 5.21 to test whether $J \cong \mathrm{Sp}(4, q)$ (cf. **3.2.3**). For each conjugate the probability of this isomorphism is $\geq 1/2^5$ (by Lemma 5.10(v)), and Corollary 5.21 succeeds with probability $\geq 1 - 1/8$. Thus, we obtain the desired type of J with probability $\geq (1/2^5)(7/8) > 1/2^6$ for each conjugate and hence with probability $> 1 - 1/16m^2$ for at least one of our $\lceil 2^7 \ln(4m) \rceil$ conjugates.

If Corollary 5.21 succeeds then we can find the 4–space V_J underlying J, the generating set $\mathcal{S}_J^*$ and the isomorphism $\lambda_J \colon J \to \mathrm{Sp}(V_J)$, and hence also the following:

(i) A transvection group T of J such that the $a\lambda_J$-invariant center α of $T\lambda_J$ is in $[V_J, a\lambda_J]$;

(ii) $Q_4 := O_p((J\lambda_L)_\alpha)\lambda_L^{-1}$, so that $|Q_4| = q^3$; and, for each generator x of Q_4, generators of the root group containing x;

(iii) G_4, the subgroup of J_α generated by its transvection groups, and a subgroup $D \cong \mathrm{Sp}(2, q)$, such that $G_4 = Q_4 \rtimes D$ and $D\lambda_J$ fixes a point γ of V_J not perpendicular to α;

(iv) $j(\gamma) \in J$ such that $j(\gamma)\lambda_J$ is a transvection moving α to γ; and

(v) $s \in J$ of order $q - 1$, such that $s\lambda_J$ fixes α and γ and induces the identity on the orthogonal complement of $\langle \alpha, \gamma \rangle$ within V_J; we can find a straight-line program of length $O(\log q)$ from $\mathcal{S}_J^*$ to any given power of s.

Increasing the set $\mathcal{S}_J^*$: We assume that $\mathcal{S}_J^*$ is increased to include generating sets of T, Q_4 and G_4 consisting of root elements, as well as the transvection $j(\gamma)$.

Reliability: $\geq 1 - 2/16m^2$.

Time: Las Vegas $O(m \log m[\xi + \mu e m^2 \log q] + \log m[\xi q e + \mu q \log^2 q])$.

Case of small q: We still need to deal with the possibility that q is small. (N.B.—In the present context this means that $q = 3$ or 5. However, we note that a slight modification of the above procedure, and of definitions in the next section, produce the desired information if we use $q = 5$ and $w = 5 + 1$.)

Choose up to $\lceil 8qm \ln(4m) \rceil$ elements $\tau \in G$ in order to find one of $p \cdot \mathrm{ppd}^\#(p; 2e(m-1))$–order. Then $t := \tau^z$ is a transvection, where $z := q^{m-1} + 1$. The probability is $> 1 - 1/16m^2$ that some τ will have the desired order (this probability is $\geq 1/4(m-1)q$ for one choice of τ, in view of **5.1.5**).

Choose up to $\lceil 24\ln(4m)\rceil$ triples t_1, t_2, t_3 of conjugates of t. For each triple let $J := \langle t, t_1, t_2, t_3 \rangle$, and use Corollary 5.21 to test whether $J \cong \mathrm{Sp}(4, q)$ (cf. **3.2.3**), in which case $\dim [V, J] = 4$. For each triple the probability of this isomorphism is $> 1/10$ (by Lemma 5.8), and Corollary 5.21 succeeds with probability $\geq 1 - 1/8$. Thus, we obtain the desired type of J with probability $\geq (1/10)(7/8) > 1/12$ for each triple and hence with probability $> 1 - 1/16m^2$ for at least one of our $\lceil 24\ln(4m)\rceil$ triples.

If Corollary 5.21 succeeds then once again we can find (i-v), where (i) provides us with the transvection group T in J containing t (and hence normalized by τ).

Reliability: $\geq 1 - 2/16m^2$.

Time: Las Vegas $O(qm\log m[\xi + \mu em^2 \log q] + \log m[\xi qe + \mu q \log^2 q])$ which, for bounded q, is $O(m\log m[\xi + \mu m^2])$.

5.2.2   Finding root groups and J: $p = 2$

Here it seems difficult to generate a subgroup of G that is isomorphic to $\mathrm{Sp}(4, q)$ and is in the desired conjugacy class of such subgroups. Instead we revert to the methodology of Section 4 since $G \cong \Omega(V') = \Omega(2m + 1, q)$.

Case $q > 2$:   Choose up to $\lceil 16m\ln(8m)\rceil$ elements of G in order to find τ of order $\mathrm{ppd}^{\#}(2; e) \cdot \mathrm{ppd}^{\#}(2; 2e(m - 1))$. Once again τ must split V as $V = V_2 \perp V_{2m-2}$, inducing elements of order dividing $w := q - 1$ on V_2 and $z := q^{m-1} + 1$ on V_{2m-2} since $(w, z) = 1$. Then $a := \tau^z$ has $\mathrm{ppd}^{\#}(2; e)$–order.

Choose up to $\lceil 2^{12}10\ln(8m)\rceil$ pairs b, c of conjugates of a, and for each pair use Section 3 to test whether $J := \langle a, b, c \rangle \cong \Omega^+(6, q)$ (here $\Omega^+(6, q) \cong \mathrm{PSL}(4, q)$). For a single choice J will be as desired with probability $\geq 1/2^{10}10$ (by Lemma 5.10(ii)), and our isomorphism test will succeed with probability $\geq 1/2$. Thus, with probability $> 1 - 1/64m^2$, for some choice of b, c we obtain the desired J together with $\lambda_J : J \to \mathrm{Sp}(V_J)$ and a generating set $\mathcal{S}_J^*$ (cf. **4.6.1**).

Case $q = 2$:   Choose up to $\lceil 8m\ln(8m)\rceil$ elements of G in order to find τ of order $5\mathrm{ppd}^{\#}(2; 2(m - 2))$ if $4 \nmid m$ and $5\mathrm{ppd}^{\#}(2; m - 2)$ if $4 \mid m$; let $z := 2^{m-2} + 1$ or $2^{m-2} - 1$, respectively. This time τ must split V as $V = V_4 \perp V_{2m-4}$, inducing elements of order dividing 5 on V_4 and z on V_{2m-4} since $(5, z) = 1$. Then $a := \tau^z$ has order 5.

Choose up to $\lceil 8\ln(8m)\rceil$ conjugates b of a, and for each use brute force to test whether $\langle a, b \rangle \cong \Omega^-(8, 2)$, in which case find an isomorphism $\lambda_J : J \to \Omega(V_J) = \Omega^-(8, 2)$. By Lemma 5.10(iv), $\langle a, b \rangle$ behaves in this manner with probability $\geq 1/4$, and hence with probability $> 1 - 1/64m^2$ for at least one of our choices b we will have $\langle a, b \rangle \cong \Omega^-(8, 2)$.

For any even q we now have an orthogonal 6– or 8–space V_J underlying J, a generating set $\mathcal{S}_J^*$, and an isomorphism $\lambda_J \colon J \to \Omega(V_J) = \Omega^+(6, q)$ or $\Omega^-(8, 2)$. As in Corollaries 3.20 and 5.21 we can use these to find the following:

(i) A long root group R of J such that $[V_J, R\lambda_J]$ contains a singular point α of $[V_J, a\lambda_J]$ that is $a\lambda_J$–invariant if $q > 2$;

(ii) $Q_6 := O_2((J\lambda_L)_\alpha)\lambda_L^{-1}$, so that $|Q_6| = q^4$ (or 2^6 if $q = 2$) and $R \leq Q_6$; a generating set for Q_6 consisting of long root elements of J, one of which is $r \in R$; and, for each such generator x, generators of the root group containing x;

(iii) G_6, the subgroup of J_α generated by its long root groups, and a subgroup $D \cong \Omega^+(4, q)$ (or $\Omega^+(6, 2)$ if $q = 2$) such that $G_6 = Q_6 \rtimes D$ and $D\lambda_J$ fixes a point γ of V_J not perpendicular to α;

(iv) $j(\gamma) \in J$ such that $j(\gamma)\lambda_J$ moves α to γ, together with a straight-line program of length $O(\log q)$ from $\mathcal{S}_J^*$ to $j(\gamma)$; and

(v) $s \in J$ of order $q - 1$ such that $s\lambda_J$ fixes α and γ and induces the identity on the orthogonal complement of $\langle \alpha, \gamma \rangle$ within V_J; we can find a straight-line program of length $O(\log q)$ from $\mathcal{S}_J^*$ to any given power of s.

Increasing the set $\mathcal{S}_J^*$: We assume that these are increased to include generating sets of R, Q_6 and G_6 consisting of long root elements.

Reliability: $> 1 - 2/64m^2$.

Time: Las Vegas $O(m \log m[\xi + \mu e m^2 \log q] + \log m[\xi qe + \mu q \log^2 q])$.

A second element τ': We will need a second element τ', conjugate to τ, having the same element $a \in \langle \tau' \rangle$, and such that τ and τ' *are independent elements of* $C_G(a)$.

Case $q \geq 8$: Choose up to $\lceil 80 \log(8m) \rceil$ conjugates τ^g, $g \in G$; test whether $\langle a, a^g \rangle \cong \Omega^+(4, q)$ by using **3.6.2**; when this isomorphism occurs use it to find $h \in \langle a, a^g \rangle$ such that $a^{gh} = a$. Let $\tau' := \tau^{gh}$.

In order to prove correctness, note that $\langle a, a^g \rangle \cong \Omega^+(4, q)$ with probability $\geq 1/20$ for a single choice a^g (by Lemma 5.10(i), since $q \geq 8$), and **3.6.2** succeeds with probability $\geq 1/2$ if this isomorphism holds. It is clear that τ and τ' are independent of one another.

Case $q = 2$: Choose up to $\lceil 16 \log(8m) \rceil$ conjugates τ^g, $g \in G$; then $\langle a, a^g \rangle \cong \Omega^-(8, 2)$ with probability $> 1/8$ for a single choice a^g by Lemma 5.10(ii). Use brute force to test this and to find $h \in \langle a, a^g \rangle$ such that $a^{gh} = a$, and let $\tau' := \tau^{gh}$.

Case $q = 4$: Choose up to $\lceil 32 \log(8m) \rceil$ conjugates τ^g, and for each use brute force to test whether $\langle a \rangle$ and $\langle a^g \rangle$ are conjugate in the group $\langle a, a^g \rangle$ of order $< 4^6$. By Lemma 5.10(iv) we will succeed with probability $> 1/16$ for a single choice τ^g. Now find τ' as above.

Reliability: $> 1 - 1/64m^2$.

Time: $O(\log m[\xi + (\xi qe + \mu q \log^2 q)])$.

5.3   Finding $Q = Q(\alpha)$, $Q(\gamma)$, G'_α, L and $\mathcal{S}^*$

5.3.1 Q, T and G'_α

Write $w := q - 1$ if $q > 5$ is odd, $w := q$ if $q = 3$ or 5, $w := q + 1$ if $q > 2$ is even, and $w := 5$ if $q = 2$. Let

$$Q := \langle Q_4^{\tau^{wi}} \mid 0 \leq i \leq 2m \rangle, \qquad G'_\alpha := \langle G_4^{\tau^{wi}} \mid 0 \leq i \leq 2m \rangle \qquad \text{if } p > 2$$

$$Q := \langle Q_6^{\tau^{wi}}, Q_6^{\tau'^{wi}} \mid 0 \leq i \leq 2m \rangle, \ G'_\alpha := \langle G_6^{\tau^{wi}}, G_6^{\tau'^{wi}} \mid 0 \leq i \leq 2m \rangle \ \text{if } p = 2.$$

Here Q is generated by root groups (cf. **5.2.1**(ii) and **5.2.2**(ii)), and we assume that generators of each of these root groups are stored. As in **3.3.1** and **4.3.1**, Q lies in 'Q' $= O_p(C_G(t))$ (cf. **5.1.3**).

Lemma 5.11 *With probability* $\geq 1 - 1/16m^2$, $Q = $ 'Q' *and* $G'_\alpha = N_G(Q)'$.

Proof. Case $p > 2$: Both for large and small q we have arranged that both G_4 and τ fix α. Also, τ acts irreducibly on $\alpha^\perp/\alpha$ and hence on 'Q'$/T$. By Lemma 2.7, the subgroups $Q_4^{\tau^{wi}}/T$, $0 \leq i \leq 2m - 2$, generate 'Q'$/T$, where $T = \Phi($'Q'$)$, so that $Q = $ 'Q'. In view of **5.1.2**, we can proceed exactly as in the proof of Lemma 4.14. As in that proof, we obtain pairwise perpendicular subspaces Q_j of Q/T, so that there are at most $(2m-2)/2$ such subspaces. On the other hand, as in that proof, τ^w permutes these Q_j. If there is more than one subspace Q_j then there is a power of τ that permutes the Q_j and has order a prime $r = \text{ppd}^\#(p; 2e(m-1))$, and hence $2m-2|r-1$, whereas $r \leq (2m-2)/2$.

Thus, there is just one Q_j, and G'_α/Q is isomorphic to an irreducible subgroup of $\mathrm{Sp}(2m-2,q)$ generated by transvection groups. By Theorem 3.6, since $2m-2 \geq 4$ and $p > 2$ the only possibility is $G'_\alpha/Q \cong \mathrm{Sp}(2m-2,q)$.

Case $p = 2 < q$: The elements τ^w and τ'^w of $C_G(a)$ fix the a–invariant singular point α of $[V_J, a]$. They have $\mathrm{ppd}^{\#}(2; 2e(m-1))$–order and are conjugate elements of $C_G(a)$ of that order that occur equally often (see the construction in **5.2.2**). There are $q^{m-1}(q^{m-1}-1)/2$ possibilities for each of the nonsingular hyperplanes $['Q', \tau^w]$ and $['Q', \tau'^w]$, and hence these are different with probability $1 - 2/q^{m-1}(q^{m-1}-1) \geq 1 - 1/16m^2$ (recall that we have assumed that $m \geq 4$ if $q = 4$). We now assume that these are, indeed, different.

By Lemma 2.7 the groups $Q_6^{\tau^{wi}}$ and $Q_6^{\tau'^{wi}}$, $0 \leq i \leq 2m-2$, generate $\langle ['Q', \tau^w], ['Q', \tau'^w] \rangle = 'Q'$. Then $'Q' = Q$, and we can proceed exactly as in the proof of Lemma 4.14 (working in $'Q'/T$, with $\dim ['Q'/T, G_6^{\tau^{wi}}] = 4$ instead of 6) in order to show that G'_α induces a subgroup of $\Omega(2m-1,q)$ generated by a conjugacy class of long root groups. Moreover, G'_α/Q cannot leave invariant any hyperplane of Q since we have generated $'Q'$, and hence by Lemma 5.9 we have $G'_\alpha/Q \cong \Omega(2m-1,q)$. (Note that $G_2(q)$ does not have a subgroup isomorphic to $\Omega^+(4,q)$ generated by long root groups of $\Omega(7,q)$.)

Case $q = 2$: Here τ^5 and τ'^5 fix the singular point α of $[V_J, a]$. They have $\mathrm{ppd}^{\#}(2; 2(m-2))$– or $\mathrm{ppd}^{\#}(2; m-2)$–order and are conjugate elements of $C_G(a)$ of that order that occur equally often (see **5.2.2**). There are $2^{3m-7}(2^{m-2} \pm 1)(2^{2m-2} - 1)/3$ possibilities for each of the nonsingular $2m-4$–spaces $['Q', \tau^5]$ and $['Q', \tau'^5]$. The first of these lies in 7 hyperplanes of $'Q'$; the second also lies in one of these with probability $< 7(2^{2m-2} - 1)3/2^{3m-7}(2^{m-2} \pm 1)(2^{2m-2} - 1) \leq 1/16m^2$ since we have assumed that $m \geq 6$. Thus, $\langle ['Q', \tau^5], ['Q', \tau'^5] \rangle = 'Q'$ with probability $\geq 1 - 1/16m^2$; we now assume that this equality holds.

This time $\dim ['Q'/T, G_6^{\tau^{5i}}] = 6$, and once again the groups $['Q', G_6^{\tau^{5i}}]$ and $['Q', G_6^{\tau'^{5i}}]$, $0 \leq i \leq 2m-2$, generate $\langle ['Q', \tau^5], ['Q', \tau'^5] \rangle = 'Q'$. Now proceed as before in order to show that $G'_\alpha/Q \cong \Omega(2m-1,2)$. $\square$

Remark. We already have the transvection group $T = \mathrm{Z}(G'_\alpha)$ for odd q. We now use G'_α to find T for even q.

Finding T when $p = 2$: Choose up to $\lceil 350 em \log(64m^2) \rceil$ elements $\sigma \in G'_\alpha$; for each test whether $|\sigma| = 2\mathrm{ppd}^{\#}(2; 2e(m-1))$, in which case let $u := \sigma^{q^{m-1}+1}$. In view of **5.1.5**, one element σ has the stated order with probability $\geq (1/4m)(q-1)/q$, and then $1 \neq u \in T = \mathrm{Z}(G'_\alpha)$. An application of Lemma 2.8 of type **2.5**(1) with the parameters $r = 1/(q+1), t = \lceil 350 em \log(64m^2) \rceil, \varepsilon = 1/4$, shows that we *get at least* $\lceil 32e \log(64m^2) \rceil$ *transvections* u *lying in* T with probability $> 1 - 1/64m^2$. We claim that *these* $\lceil 32e \log(64m^2) \rceil$ *transvections* $u_1, \ldots, u_{\lceil 32e \log(64m^2) \rceil}$ *generate* T *with probability* $> 1 - 1/64m^2$. For, if $\langle u_1, \ldots, u_{i-1} \rangle \neq T$ then $\text{Prob}(u_i \notin \langle u_1, \ldots, u_{i-1} \rangle) \geq ((2^e - 1) - 2^{e-1})/(2^e - 1) > 1/8$. Hence an application of Lemma 2.8 of type **2.5**(2) with the parameters $r = 1/8, t = \lceil 32e \log(64m^2) \rceil, \varepsilon = 7/8$, shows that the u_i's generate T with the stated probability. Use Lemma 2.1 to test the order of the group generated, and to find a basis $\{t_1, \ldots, t_e\}$ of T.

Reliability: $> 1 - 2/64m^2$.

Time: Las Vegas $O(me \log m[\xi + \mu em^2 \log q] + \mu qe)$.

5.3.2 Effective transitivity of Q; the complement L

Points are defined to be the conjugates $T(\beta) = T^g$, $g \in G$. (We could have used conjugates of Q as in Section 4, but this definition is equivalent to that one and is much easier to work with.) Points $T(\beta)$ and $T(\gamma)$ are called *perpendicular* if and only if $[T(\beta), T(\gamma)] = 1$ (which can be tested by checking just one generator of $T(\beta)$ and one of $T(\gamma)$). We may assume that the elements of T have already been listed. We have $1 \neq t \in T$.

Lemma 5.12 *In deterministic* $O(\mu[qe+m])$ *time an element of* Q *can be found conjugating any given point* T^x *not perpendicular to* T *to any other given point* T^y *not perpendicular to* T.

Proof. Either $[t^x, t^y] \neq 1$ or $[t^x, t^{yt}] \neq 1$; since $t \in Q$ we may assume that $[t^x, t^y] \neq 1$.

Test each $u \in (t^y)^{T^x}$ in order to find one commuting with t. Test a single generator in each of the $O(m)$ root groups R generating Q in order to find one such that $[R, u] \neq 1$. If no such root group is found then test all $v \in T$ to find one satisfying $[t^{xv}, t^y] = 1$, and output v.

Let $A := [R, u]T < Q$. If $q \leq 3$ find $v \in A$ such that $t^{xv} \in T^y$, and output v. For large q we wish to avoid searching through the q^2 elements of A.

Find $t' \in T$ such that $[(t^y)^{t'}, t^x] = 1$. Let $z := yt'$. It suffices to find an element of Q conjugating T^x to T^z.

For each generator $a \in A$ find $a' \in T$ such that $[t^{xaa'}, t^x] = 1$. Let B be the group (of order q) generated by the elements aa'. List B. (The desired element is in B, but we need a test in order to determine this element.)

Let $t_1 := t$ and $1 \neq t_2 \in T - \{t_1, t_1^{-1}\}$ (recall that $q \geq 4$). For $i = 1, 2$ find $1 \neq s_i \in T^x$ such that $[t, (t^{xt_i})^{s_i}] = 1$. Let $g := t_1 s_1 t_1 t_2 s_2 t_2 \in \langle T, T^x \rangle$; this normalizes T and T^x but does not centralize either of them. Similarly obtain $h \in \langle T, T^z \rangle$ normalizing T and T^z but not centralizing either of them.

Find $b \in B$ such that $[g^b, h] = 1$. Output b.

Correctness: T^x and T^y correspond to points $\gamma, \delta \notin \alpha^\perp$. By (5.1), the points α, δ, δ^t are collinear, and there is just one point of their line lying in $\gamma^\perp$. Thus, either $\delta \notin \gamma^\perp$ or $\delta^t \notin \gamma^\perp$; we may assume that $\delta \notin \gamma^\perp$, so that $\langle T^x, T^y \rangle \cong \mathrm{SL}(2, q)$ by **5.1.2**.

If $\alpha \in \langle \gamma, \delta \rangle$ then some $v \in T$ sends γ to δ. Suppose that $\langle \alpha, \gamma, \delta \rangle$ has dimension 3 and let $\varepsilon = \alpha^\perp \cap \langle \gamma, \delta \rangle$. Since T^x is transitive on the points $\neq \langle \gamma \rangle$ of $\langle \gamma, \delta \rangle$, it has an element w sending δ to ε. Then $u = (t^y)^w \in T(\varepsilon)$, so that u commutes with t and hence normalizes Q; no other element of $(t^y)^{T^x}$ commutes with t.

Note that u induces a transvection on Q/T with center $[Q, u]T/T$ corresponding to $\langle \alpha, \varepsilon \rangle / \alpha$ (cf. **5.1.3**; note that $T \leq [Q, u]$ if and only if q is odd). Thus, $[R, u]T = [Q, u]T$. Each element of $[R, u]$ induces a transvection on $\langle \alpha, \gamma, \delta \rangle$ having axis $\langle \alpha, \varepsilon \rangle$, and A contains all q^2 such transvections. In particular, some element of this group sends γ to δ. Thus, the algorithm is correct if $q \leq 3$.

We found $\alpha^z = \langle \alpha, \alpha^y \rangle \cap (\alpha^x)^\perp$. Then T^x and T^z commute, and generate the group of all transvections of $\langle \alpha, \gamma, \alpha^z \rangle$ with axis $\langle \alpha^x, \alpha^z \rangle$; the only elements of $(t^x)^A$ commuting with t^x lie in this group and hence fix $\langle \alpha^x, \alpha^z \rangle$. Thus, B consists of transvections of $\langle \alpha, \gamma, \alpha^z \rangle$ with center $\zeta = \langle \alpha^x, \alpha^z \rangle \cap \alpha^\perp$, one of which moves $\gamma = \alpha^x$ to α^z.

We found $h \in \langle T, T^z \rangle \cong \mathrm{SL}(2, q)$ by using the identity

$$\begin{pmatrix} 1 & k \\ 0 & 1 \end{pmatrix} \begin{pmatrix} 1 & 0 \\ -k^{-1} & 1 \end{pmatrix} \begin{pmatrix} 1 & k \\ 0 & 1 \end{pmatrix} = \begin{pmatrix} 0 & k \\ -k^{-1} & 0 \end{pmatrix}$$

for two elements k of $\mathbb{F}$: h is in the cyclic group $N_z = \mathrm{N}_{\langle T, T^z \rangle}(T) \cap \mathrm{N}_{\langle T, T^z \rangle}(T^z)$ of order $q - 1$, but $|h| > 2$. The only points of $\langle \alpha^x, \alpha^z \rangle$ fixed by h are ζ and α^z, while the only ones fixed by g are ζ and α^x. Thus, $b \in B$ sends α^x to α^z if and only if it conjugates g to an element of N_z, which occurs if and only if g^b fixes α and α^z, i.e., if and only if g^b commutes with h.

Time: $O(\mu[qe + m])$, dominated by finding R, and finding and then listing the elementary abelian group B using Lemma 2.1. $\square$

Remarks. 1. A simpler Las Vegas $O(\xi q + \mu q \log^2 q)$ time algorithm for the lemma is as follows. Find $t' \in T$ such that $[t^{xt'}, t^y] = 1$, and replace t^x by $t^{xt'}$.

As in **3.6.1** (probably) find $g \in \langle T, T^x \rangle$ of order $q - 1$ normalizing T and T^x, and $h \in \langle T, T^y \rangle$ of order $q - 1$ normalizing T and T^y. Let $v := [h, g]$. Find $0 \le i \le q - 2$ such that $[h, g^{v^{g^i}}] = 1$. Output v^{g^i}.

2. Let $Q(\gamma) := Q^{j(\gamma)}$ and $T(\gamma) := T^{j(\gamma)}$.

By (5.5), the element of Q found in the lemma is *unique*, and there is a complement to Q in G'_α. One can be found as in the proof of Corollary 4.18:

Corollary 5.13 *In deterministic $O(\mu[qem + m^2])$ time a complement $L \cong G'_\alpha/Q \cong \mathrm{Sp}(2m - 2, q)$ to Q in G'_α can be found centralizing $T(\gamma)$.*

5.3.3   Recursion: S^* and λ_L

Once again we will require that *our set S^* of generators of G consists entirely of root elements.* Using Theorem 1.1', recursively test whether $L \cong \mathrm{Sp}(2m-2, q)$; by Lemma 5.11, this holds with probability $> 1 - 1/16m^2$, while the test verifies this with certainty. As in **3.3.3** and **4.3.3**, the recursive call provides *a set S_L^* of generators of L, a vector space V_L, and an isomorphism $\lambda_L \colon L \to \mathrm{Sp}(V_L) = \mathrm{Sp}(2m - 2, q)$ defined on S_L^*.* Then $G = \langle G'_\alpha, J \rangle$ is generated by the set

$$S^* = S^*(1) \cup S^*(2) \cup S^*(3)$$

of $O(em^2)$ root elements, where $S^*(1) = S_L^*$, $S^*(2)$ is the defining set of $O(em)$ generators for Q (cf. **5.3.1**) together with generators for T, and $S^*(3) = S_J^*$ is our set of $O(e)$ generators for J.

Computing preimages recursively: Once again, as in **3.3.3** we will need to be able to apply λ_L^{-1} to elements of $L\lambda_L$. This is the content of Proposition 5.17 (and of Theorems 1.1(iv) and 1.1'(iv)), with L in place of G: *in $O(\mu m^2 \log q)$ time, $h\lambda_L^{-1}$ can be found for any given $h \in L\lambda_L$.*

The choice of λ_L when $G/Z(G) \cong \mathrm{PSp}(6, q)$ with q even: Recursion is slightly less straightforward here: whereas **5.6.1**, or even just brute force, produces a 4–dimensional vector space for L, there are *two* vector spaces that might arise (permuted transitively by $\mathrm{Aut}(L)$). This situation is analogous to the one in **3.3.3**, where the possibility of two equally natural L–modules had to be considered: as in **3.3.3** we may need to replace λ_L by $\lambda_L\iota$, where ι denotes the outer automorphism of $L\lambda_L$ constructed in **5.6.3**. Both λ_L and $\lambda_L\iota$ produce isomorphisms $L \to \mathrm{Sp}(4, q)$, but only one of them will give the action needed: we need to know that, when we fix a nonzero vector using $L\lambda_L$, the preimage in L will also fix a nontrivial element of Q/T. We achieve this as follows.

In the derived subgroup P_0 of the stabilizer in $L\lambda_L$ of a 1-space find the normal subgroup Q_0 of order q^3 (cf. **5.1.3**). Test whether $[[Q, Q_0\lambda_L^{-1}], P_0\lambda_L^{-1}] \neq 1$; if so then replace λ_L by $\lambda_L\iota$.

Correctness: If $[Q, R\lambda_L^{-1}]T/T$ is a 2–space of Q/T then $P_0\lambda_L^{-1}$ acts nontrivially on it; but if $[Q, R\lambda_L^{-1}]T/T$ is a 1–space then $P_0\lambda_L^{-1}$ centralizes it and hence centralizes $[Q, R\lambda_L^{-1}]$ (since $P_0\lambda_L^{-1}$ has no normal subgroup of index p).

Time: $O(\mu e \log q)$, using Proposition 5.17 to apply λ_L^{-1}: we may assume that P_0 has $O(1)$ generators.

Two generators: It is straightforward to write down 2 generators (or, if preferred, $O(1)$ generators) of $L\lambda_L$, chosen as a matter of convenience. Use Proposition 5.17 to pull them back to generators of L.

The isomorphism $\lambda^{\#}$: Throughout this section we are assuming that there exists an isomorphism $\lambda^{\#} \colon G \to \mathrm{Sp}(V)$ or $\mathrm{PSp}(V)$. While the goal is to construct such an isomorphism, as in **3.3.3** we use its existence to motivate and prove correctness of subroutines in the algorithm. Recall the convention in **3.3.1** that, for example, "Q" denotes the matrix group studied in **5.1.3**. Clearly $\lambda^{\#}$ sends the nonperpendicular points $Q, Q(\gamma)$ to a nonperpendicular pair of points of V, which we may assume is α, γ in the notation of **5.1.3**. As in **3.4.4** and **4.4.4** we can use $\lambda^{\#}$ to see that λ_L extends to an isomorphism $\lambda^{\bullet} \colon G'_\alpha = QL \to$ "Q""L" determined by a semilinear map $Q \to$ "Q", and $\lambda^{\bullet}|_Q$ also preserves the form on Q/T in the sense that, for some $k \in \mathbb{F}^*$ and $\sigma \in \mathrm{Aut}\,\mathbb{F}$, we have $(uT\lambda^{\bullet}, vT\lambda^{\bullet})'' = k(uT, vT)_Q^\sigma$ for all $u, v \in Q$, where $(\ ,\ )''$ and $(\ ,\ )_Q$ denote the alternating forms on "Q"/"T" and Q/T, respectively.

By following $\lambda^{\#}$ with a field automorphism we see that there is an isomorphism $\lambda^{\bullet}$ of the desired sort such that the map $\lambda^{\bullet}|_Q$ is linear on Q/T. In Lemma 5.16 we will find $\lambda^{\bullet}|_Q$. For now, we note only that $\lambda^{\bullet}|_Q$ is uniquely determined up to a scalar transformation.

5.4 Deterministic algorithms for Q

All procedures in 5.4 and 5.5 will be deterministic.

5.4.1 Decomposition of Q/T

Lemma 5.14 *In deterministic $O(\mu m^2 \log q)$ time $O(e)$–generator subgroups $A_i \leq Q$, $i \in \{\pm 1, \ldots, \pm(m-1)\}$, can be found such that the following hold:*

(i) *$|A_i| = q$ and Q/T is the direct product of the groups $A_i T/T$; and*

(ii) *In deterministic $O(\mu m)$ time, any given $uT \in Q/T$ can be expressed in the form $uT = \prod_{1 \leq |i| \leq m-1} a_i T$ with $a_i \in A_i$.*

Proof. We have a representation λ_L of L on the space $V_L = \mathbb{F}^{2m-2}$ equipped with a nonsingular alternating bilinear form. We may assume that the standard basis of $\mathbb{F}^{2m-2}$ is $e_1, e_{-1}, \ldots, e_{m-1}, e_{-(m-1)}$, and that the form is given so that this is a standard basis in the symplectic sense as well: $V_L = W_1 \perp \cdots \perp W_{m-1}$ with $W_i = \langle e_i, e_{-i} \rangle$ and $(e_i, e_{-i}) = 1$ for each i. Let $c' \in \mathrm{Sp}(2m-2, q)$ send $e_1 \mapsto \cdots \mapsto e_{m-1} \mapsto e_1$ and $e_{-1} \mapsto \cdots \mapsto e_{-(m-1)} \mapsto e_{-1}$. Let $j' \in \mathrm{Sp}(2m-2, q)$ send $e_i \mapsto e_{-i} \mapsto -e_i$ for $1 \leq i \leq m-1$. In notation corresponding to that in the proof of Lemma 3.13, both $t_1' := I + E_{-1,1}$ and $t_{-1}' := I - E_{1,-1}$ belong to $L\lambda_L$. Also, let $h_1' := \mathrm{diag}(\rho, \rho^{-1}, 1, \ldots, 1)$. Use Proposition 5.17 to find $c := c' \lambda_L^{-1}$, $j := j' \lambda_L^{-1}$, $h_1 := h_1' \lambda_L^{-1}$ and $t_1 := t_1' \lambda_L^{-1}$. Let $t_{-1} := t_1^{-j}$, $t_{\pm i} := t_{\pm 1}^{c^{i-1}}$ and $h_i := h_1^{c^{i-1}}$ for $2 \leq i \leq m-1$.

Find a generator r_1 of Q such that $[r_1, t_1] \neq 1$; let $R_1 < Q$ be the root group containing r_1 (cf. **5.3.1**). By (5.1) and (5.4), $[Q, t_1] = [R_1, t_1]$ if $p = 2$, while $[Q, t_1] = [R_1, t_1]T$ if $p > 2$; in either case, we obtain $O(e)$ generators for $[Q, t_1]$. Let

$$
\begin{aligned}
A_1 &:= [[Q, t_1], h_1] && \text{if } q > 2. \\
A_1 &:= \langle q_1 t \rangle && \text{if } q = 2, \text{ where } [Q, t_1] = \langle q_1 \rangle \\
A_{-1} &:= A_1^j \\
A_{\pm i} &:= A_{\pm 1}^{c^{i-1}} && \text{for } 2 \leq i \leq m-1.
\end{aligned}
\tag{5.15}
$$

Then Q/T is the direct product of the groups $A_{\pm i} T/T$, and $|A_{\pm i}| = q$ since h_1 centralizes the subgroup T of $[Q, t_1]$ for any p. (Note that each $A_{\pm i}$ is a short root group.)

Given $uT \in Q$, we have $uT = \prod_1^{m-1}(a_i a_{-i})T$ where

$$a_{\pm i} := [[u, t_{\mp i}], h_i^{\mp 1}]^{j^{\mp 1} h_i^{\pm n}} \qquad \text{if } q > 2$$

$$a_{\pm i} := \begin{cases} 1 & \text{if } [u, t_{\mp i}] = 1 \\ [u, t_{\mp i}]^j t & \text{otherwise} \end{cases} \qquad \text{if } q = 2$$

for an integer n satisfying $\rho^n = (\rho - 1)^{-1}$.

Time: $O(\mu m^2 \log q)$ to find the subgroups $A_{\pm i}$, dominated by applying λ_L^{-1} four times using Proposition 5.17.

Correctness: First assume that $q > 2$. Let $v = r(z, l) \in$ "Q" in the notation of (5.3), where $z = \Sigma_i k_i e_i$. By (5.7) and (5.4) we have $[[v, t'_{\mp i}], h_i'^{\mp 1}] = r((\rho - 1)k_{\pm i}e_{\mp i}, 0)$ for $1 \leq i \leq m - 1$, where $h_i' = h_1'^{c'^{i-1}}$. It follows that $\prod_i[[v, t'_{\mp i}], h_i'^{\mp 1}]^{j'^{\mp 1} h_i'^{\pm n}} \equiv \prod_i r(k_{\pm i}e_{\pm i}, 0) \equiv r(z, l) \pmod{\text{"}T\text{"}}$. At the end of **5.3.3** we saw that there is an isomorphism $\lambda^\bullet : QL \to$ "Q" "L" extending λ_L. Then

$$\prod_i [[v\lambda^{\bullet -1}, t_{\mp i}], h_i^{\mp 1}]^{j^{\mp 1} h_i^{\pm n}} \equiv v\lambda^{\bullet -1} \pmod{T}$$

with

$$[[v\lambda^{\bullet -1}, t_{\mp i}], h_i^{\mp 1}]^{j^{\mp 1} h_i^{\pm n}} \in [[Q, t_{\mp i}], h_i^{\mp 1}]^{j^{\mp 1}} T/T = A_{\pm i}T/T$$

for an arbitrary element $u = v\lambda^{\bullet -1}$ of Q.

Similarly, if $q = 2$ then $[v, t'_{\mp i}] = r(k_{\pm i}e_{\mp i}, k_{\pm i}) = 1$ or $r(e_{\mp i}, 1)$ for $1 \leq i \leq m - 1$. Now apply $\lambda^{\bullet -1}$ as above. $\square$

5.4.2   A field of endomorphisms of Q/T

By (5.6), the element s in **5.2.1**(v) or **5.2.2**(v) acts on Q/T as an automorphism generating the multiplicative group of the desired field $\mathrm{GF}(q)$.

Note that this definition has the additional advantage that we obtain an *action of* $\mathrm{GF}(q)$ *on* Q, not just on Q/T.

Identification with $\mathbb{F}$.   We need to relate this field to the field $\mathbb{F}$ already available from L. In the proof of Lemma 5.14 we found $h_1' \in L\lambda_L$ and $h_1 = h_1' \lambda_L^{-1} \in L$, where h_1' induces ρ on W_1. Let $1 \neq b_1 \in A_1 T/T$, and find the power s^i such that $b_1^{s^i} = b_1^{h_1}$. Now identify s^i with ρ.

5.4.3 Q/T as an $\mathbb{F}$–space

(1) *Bases $\mathcal{B}$ and $\mathcal{B}_p$:* We have a basis $e_{\pm i}$, $1 \leq i \leq m - 1$, of V_L, where $e_{\pm i} = e_1^{l'(i)}$ with $l'(i) = c'^{i-1}$, $l'(-i) = j'c'^{i-1}$ for $1 \leq i \leq m - 1$. We chose

b_1 in **5.4.2**. Let $b_i := b_1^{l(i)}$ for $1 \le |i| \le m - 1$, where $l(i) := l'(i)\lambda_L^{-1} = c^{i-1}$ or jc^{i-1}. Then $\mathcal{B} := \{b_{\pm i} \mid 1 \le i \le m - 1\}$ is an $\mathbb{F}$–basis for Q/T, while $\mathcal{B}_p := \{b^{s^j} \mid b \in \mathcal{B}, 0 \le j < e\}$ is a GF(p)–basis for Q/T. For $1 \le |i| \le m - 1$, list A_i as $\{1\} \cup \{b_i^{s^j} \mid 0 \le j < q - 1\}$. **Time:** $O(\mu q m)$.

(2) *Linear combinations.* Once again, we can use the GF(p)-linear combinations stored with $\mathbb{F}$ in **2.3** to switch between linear combinations of $\mathcal{B}$ and of $\mathcal{B}_p$.

Time: $O(\mu q m)$ to write any $u \in Q/T$ using $\mathcal{B}$ or $\mathcal{B}_p$, or to obtain u by using a straight-line program from $\mathcal{B}_p$ of length $O(m \log q)$.

(3) *Modify $\mathcal{S}^*(2)$:* We may assume that $\mathcal{S}^*(2)$ contains preimages in $\bigcup_i A_i$ of all members of $\mathcal{B}_p$, together with generators for the specific transvection group T obtained in **5.2.1** or **5.3.1**.

(4) *Matrices $\tilde{g}$:* Using $\mathcal{B}$ and (2) we can find the matrix $\tilde{g}$ representing the linear transformation induced on the $\mathbb{F}$–space Q/T by any given $g \in N_G(T)$. **Time:** $O(m \cdot \mu q m)$.

5.4.4   The extension $\lambda^\bullet$

As in **3.4.4** and **4.4.4**, we now extend λ_L to G'_α. This time more care is needed, since Q need not be abelian. Let "Q" and "T" denote the matrix groups in **5.1.3**. Recall that, in the notation of (5.3), we have a bijection $r: \langle \alpha, \gamma \rangle^\perp \times \mathbb{F} \to$ "Q" inducing an isometry $\langle \alpha, \gamma \rangle^\perp \to$ "Q"/"T". Let "L" $\cong$ Sp$(2m - 2, q)$ denote the centralizer of $\langle \alpha, \gamma \rangle$ in the symplectic group "G" implicit in **5.1.3**.

We identify V_L with the $2m - 2$–space $\langle \alpha, \gamma \rangle^\perp$ (in any manner) , so that we now have $r: V_L \times \mathbb{F} \to$ "Q". Recall the bases $e_{\pm 1}, \ldots, e_{\pm(m-1)}$ of V_L and $b_{\pm 1}, \ldots, b_{\pm(m-1)}$ of Q/T in **5.4.3**(1).

Lemma 5.16 *In $O(\mu e m)$ time λ_L can be extended to an isomorphism $\lambda^\bullet: G'_\alpha = QL \to$ "Q" "L" with the following properties:*

(i) *if $b \in Q$ with $bT = \Sigma_i k_i b_i$ (for $k_i \in \mathbb{F}$ found in $O(\mu q m)$ time), then $b\lambda^\bullet = r(\Sigma_i k_i e_i, \rho_b)$ for some $\rho_b \in \mathbb{F}$ found in $O(\mu q)$ time; $\lambda^\bullet: Q/T \to$ "Q"/"T" is $\mathbb{F}$–linear and determines an L–invariant alternating bilinear form $(,)_{Q/T}$ on Q/T such that $\lambda^\bullet|_{Q/T}$ becomes an isometry; and*

(ii) *$\tilde{l} = l\lambda^\bullet = l\lambda_L$ whenever $l \in L$.*

Proof. At the end of **5.3.3** we saw that λ_L extends to an isomorphism $\lambda^\bullet \colon QL \to$ "Q""L" induced by a semilinear map $\lambda^\bullet|_{Q/T}$ that preserves the forms (up to a scalar and a field automorphism) and is uniquely determined up to a scalar transformation. We claim that the definition

$$b_i\lambda^\bullet := r(e_i, \mathbb{F}) \text{ for } 1 \le |i| \le m-1.$$

of $\lambda^\bullet$ on Q/T is forced by the known actions of L and $L\lambda_L$ (up to the afore-mentioned scalar ambiguity). For, since $[Q, t_1]\lambda^\bullet = [$"$Q$"$, t_1']$ in **5.4.1**, we must have $b_1\lambda^\bullet \in r(\langle e_1 \rangle, \mathbb{F})$, and then we may assume that $b_1\lambda^\bullet = r(e_1, \mathbb{F})$. Use the elements $l'(i) \in L\lambda_L$ and $l(i) = l'(i)\lambda_L^{-1} \in L$ introduced in **5.4.3**(1) in order to deduce that we must have $b_i\lambda^\bullet = (b_1^{l(i)})\lambda^\bullet = r(e_1, \mathbb{F})^{l'(i)} = r(e_i, \mathbb{F})$ for each i. Also, as in the proof of Lemma 4.20 we can use the elements h_1 and h_1' in **5.4.1** in order to see that $\lambda^\bullet$ is linear on Q/T: $(\Sigma_i k_i b_i)\lambda^\bullet = r(\Sigma_i k_i e_i, \mathbb{F})$ for all scalars k_i. Now (ii) follows as in **3.4.4**.

Moreover, this produces an isomorphism $\lambda^\bullet \colon QL/T \to$ "Q""L"$/$"T". We extend this to an isomorphism $\lambda^\bullet$ defined on $QL\langle s \rangle/T$. Namely, conjugation by s on Q was identified with scalar multiplication by ρ on Q/T in **5.4.2**, while in (5.6) there is an element "s" $\in$ "G" inducing scalar multiplication by ρ on "Q"$/$"T". Hence, $\lambda^\bullet$ extends to an isomorphism $QL\langle s \rangle/T \to$ "Q""L"$\langle$"s"$\rangle/$"T" using $s\lambda^\bullet = $"$s$"; this is exactly how $\lambda^\#$ must behave (up to a scalar; cf. **5.3.3**).

Let $\hat{b}_{\pm i} \in A_{\pm i}$ with $\hat{b}_{\pm i}T = b_{\pm i}$ for $1 \le i \le m-1$. Define $\lambda^\bullet$ *on* Q and $t(l) \in T$, $l \in \mathbb{F}$, as follows: $t(0) := 1$, and, for $0 \le k < q-1$ and $1 \le i \le m-1$,

$$\begin{aligned} t(-2\rho^k) &:= [\hat{b}_1^{s^k}, \hat{b}_{-1}] && \text{if } p > 2 \\ t(\rho^{2k}) &:= t^{s^k} && \text{if } p = 2 \\ \hat{b}_{\pm i}^{s^k}\lambda^\bullet &:= r(\rho^k e_{\pm i}, 0) \\ t(l)\lambda^\bullet &:= r(0, l). \end{aligned}$$

We claim that these definitions are essentially forced by the existence of $\lambda^\bullet$. First note that the desired $\lambda^\bullet$ must satisfy $A_1\lambda^\bullet = r(\langle e_1 \rangle, 0)$. (For, by (5.15) we have $[Q, t_1] = [[Q, t_1], h_1]$ if $q > 2$, while $[Q, t_1] = [[Q, t_1], t_{-1}]^j$ when q is even. Now apply $\lambda^\#$ in order to obain the 0 in $r(\langle e_1 \rangle, 0)$.) Then also $A_{\pm i}\lambda^\bullet = A_1^{l(\pm i)}\lambda^\bullet = r(\langle e_{\pm i} \rangle, 0)$. Thus, whenever $1 \le i \le m-1$ the desired image $\hat{b}_{\pm i}\lambda^\bullet$ must project onto $r(e_{\pm i}, \mathbb{F})$ modulo "T" and lie in $r(\langle e_{\pm i} \rangle, 0)$, so that $\hat{b}_{\pm i}\lambda^\bullet = r(e_{\pm i}, 0)$. Now we also must have $\hat{b}_i^{s^k}\lambda^\bullet = (\hat{b}_i\lambda^\bullet)^{(s\lambda^\bullet)^k} = r(e_i, 0)^{\text{"}s\text{"}^k} = r(\rho^k e_i, 0)$ using our extension of $\lambda^\bullet$ to $QL\langle s \rangle$ and the known action of "s" (cf. (5.6)).

Thus, we have reconstructed the isomorphism $\lambda^{\#}|_{QL}$: it is, indeed, $\lambda^{\bullet}$ as defined above. Since we know that the forms on Q/T and "Q"/"T" must be preserved up to a scalar and a field automorphism, while $\lambda^{\bullet}$ is linear on Q/T, we can use $\lambda^{\bullet}$ to determine the desired form on Q/T.

Finally, our definition of $\lambda^{\bullet}$ on $G'_{\alpha} = QL$ is effective on Q. Namely, let $g \in Q$. Use **5.4.3**(2) to write $gT = xT$ for a product x of powers of the $\hat{b}_i^{s^k}$, for which $\lambda^{\bullet}$–images are known. Find gx^{-1}, and write it as $t(l) \in T$ (recall that we have listed T). Then $g\lambda^{\bullet} = r(0,l)(x\lambda^{\bullet})$, as required for (i).

Time: For $\lambda^{\bullet}$ this is dominated by the time $O(\mu em + em \cdot m^2)$ to write all of the matrices $\hat{b}_i^{s^k}\lambda^{\bullet}$; finding $g\lambda^{\bullet}$ uses **5.4.3**(2). $\square$

Remark. The spaces Q/T and V_L are related via $u = \sum_i k_i b_i \mapsto \dot{u} = \sum_i k_i e_i$, so that
$$b\lambda^{\bullet} = r(\dot{bT}, \rho_b) \text{ for all } b \in Q \text{ and some } \rho_b \in \mathbb{F}.$$

5.5   The homomorphism λ and straight-line programs

Points were defined in **5.3.2**.

5.5.1   Labeling points

We now label any given *point by a 1–space in $V := \mathbb{F}^{2m}$ in a G'_{α}–invariant manner.*

Label T by the 1–space $\langle 0, \ldots, 1, 0 \rangle$ and $T(\gamma) = T^{j(\gamma)}$ by $\langle 0, \ldots, 0, 1 \rangle$.

Consider any point $T(\delta) \neq T(\gamma)$ not perpendicular to T. Use Lemma 5.12 to find the unique $b \in Q$ conjugating $T(\gamma)$ to $T(\delta)$. If $b\lambda^{\bullet} = r(\dot{bT}, \rho_b)$ (found using Lemma 5.16(i)), label $T(\delta)$ as $\langle \dot{bT}, \rho_b, 1 \rangle$. This conforms to (5.3).

Next consider a point $T(\beta) = T^g \neq T$ perpendicular to T. In **5.3.3** we found two generators of L; find one of these generators h such that $t^{g^{-1}h}$ does not commute with $t^{g^{-1}}$ although it commutes with t (note that L fixes α but moves $\alpha^{\perp} \cap (\alpha^{g^{-1}})^{\perp}$). Some generator $u \in Q$ moves $T^{g^{-1}h}$ (hence $T^{g^{-1}hu}$ corresponds to a third point of the line $\langle \alpha, \alpha^{g^{-1}h} \rangle$; then $\beta = \alpha^g, \alpha^{g^{-1}hg}, \alpha^{g^{-1}hug}$ are distinct and collinear, and the last two are not in $\alpha^{\perp}$). If $T(\alpha^{g^{-1}hg})$ and $T(\alpha^{g^{-1}hug})$ are labeled $\langle x_1, \rho_1, 1 \rangle$ and $\langle x_2, \rho_2, 1 \rangle$, respectively, then label $T(\beta)$ as $\langle x_1 - x_2, \rho_1 - \rho_2, 0 \rangle$. It is straightforward to check that this labeling is forced by that of the points not perpendicular to T.

Time: $O(\mu[qe + qm])$ to label one point: $O(\mu[qe + m])$ to find b, $O(\mu qm)$ to find $b\lambda^{\bullet}$, and $O(\mu)$ to test all generators h.

5.5.2   The homomorphism and the form

We now construct a homomorphism λ from G onto $\mathrm{PSp}(V)$.

In **5.4.3**(1) we have a basis $\mathcal{B} = \{b_i, b_{-i} \mid 1 \leq i \leq m-1\}$ of Q/T and elements $l(i) \in L$ with $b_1^{l(i)} = b_i$. By **5.4.4** and **5.5.1**, the label of $T(\gamma)^{\hat{b}_i}$ is $\langle v_i \rangle$ with $v_i := (e_i, \rho_{\hat{b}_i}, 1)$. We have already labeled T and $T(\gamma)$ as $\langle v_m \rangle$ and $\langle v_{-m} \rangle$, where $v_m := (O, 1, 0)$ and $v_{-m} := (O, 0, 1)$.

Let b_0 be the product of the elements $\hat{b}_i$ in any order, and let $b_* := c_1 c_{-1}$, where $c_i \in \hat{b}_i T$ is such that $c_i \lambda^{\bullet} = r(e_i, 0)$. Then the label of $Q(\gamma)^{b_0}$ is $\langle v_0 \rangle$, where $v_0 := (b_0 T, \rho_{b_0}, 1)$, while that of $Q(\gamma)^{b_*}$ is $\langle v_* \rangle$, where $v_* := (e_1 + e_{-1}, -1, 1)$ since $r(e_1, 0) r(e_{-1}, 0) = r(e_1 + e_{-1}, -1)$ by (5.4) (recall that $(e_1, e_{-1}) = 1$ in **5.4.1**).

Given $x \in G$ we now obtain a matrix $x\lambda$, uniquely determined up to a scalar, exactly as in **4.5.2**.

Consistency of λ and λ_L. Equation (4.21) again holds modulo scalars for any $x \in L$. Namely, let $b \in Q$. Then $b^x T = (bT)^x = (bT)(\tilde{x}) = b(x\lambda_L)$ by Lemma 5.16(i), **5.5.1** and Lemma 5.16 together with $(Q(\gamma)^b)^x = Q(\gamma)^{b^x}$ yield $\langle bT, \rho_b, 1 \rangle (x\lambda) = \langle b^x T, \rho_{b^x}, 1 \rangle = \langle bT(x\lambda_L), \rho_{b^x}, 1 \rangle$. Since $L\lambda$ must fix $\langle v_{2m-1} \rangle$ and $\langle v_{2m} \rangle$, this proves our assertion.

Once again, modulo scalars λ and $\lambda^{\bullet}$ coincide on Q. The proof is the same as in **4.5.2**: $\langle \dot{c}T, \rho_c, 1 \rangle (b\lambda) = \langle \dot{c}T + \dot{b}T, \rho_c + \rho_b - (\dot{c}T, \dot{b}T)_Q, 1 \rangle$ and $(\dot{c}T, \rho_c, 1)(b\lambda^{\bullet}) = (\dot{c}T + \dot{b}T, \rho_{cb}, 1)$, so that (5.4) implies the assertion.

Recursively finding $\mathcal{S}^*\lambda$: As in **3.5.2**, since (4.21) holds we know that $\mathcal{S}^*(1)\lambda = \mathcal{S}_L^*\lambda$ can be found from $S_L^*\lambda_L$ by bordering; we just saw that the $O(em)$ elements of $\mathcal{S}^*(2)\lambda$ are found in $O(em \cdot \mu qm)$ time using $\lambda^{\bullet}$; and $\mathcal{S}^*(3)\lambda$ can be found using e applications of **5.5.1**. **Time:** $O(em\mu[qe + qm])$ to find $\mathcal{S}^*\lambda$ recursively, in addition to the time to find $\mathcal{S}_L^*$.

Finding $\mathcal{S}^*\lambda$ in Theorem 1.1′: As in **3.5.2**, raise matrices to the $1-q$ power in order to find the exact image $x\lambda$ of all $x \in \mathcal{S}^*$. **Time:** $O(m^3 \cdot m^2 e \log q)$.

Finding an alternating bilinear form on V: This is found as in **4.5.2**.

5.5.3   Theorems 1.1(iii,iv) and 1.1′(iii,iv)

As in **3.5.3** we have

Proposition 5.17 *In deterministic $O(m^4 \log q)$ time, given $h \in \mathrm{GL}(2m, q)$,*

(a) *it can be determined whether or not $h \in G\lambda$ (modulo scalars); and*

(b) *if so then a straight-line program of length $O(m^2 \log q)$ can be found from $\mathcal{S}^*\lambda$ to h (only modulo scalars in the situation of Theorem 1.1(iv)), after which $h\lambda^{-1}$ can be found in additional $O(\mu m^2 \log q)$ time.*

Moreover,

(c) *in Theorem 1.1', $Z(G)$ can be found in $O(\mu m^2 \log q)$ time.*

Proposition 5.18 *In deterministic $O(\mu[qe + qm^2])$ time a straight-line program of length $O(m^2 \log q)$ can be found from $\mathcal{S}^*$ to any given $g \in G$. (Hence, the corresponding straight-line program can be written from $\mathcal{S}^*\lambda$ to $g\lambda$, and $g\lambda$ can be computed in additional $O(m^3 \cdot m^2 \log q)$ time.)*

The proofs are similar to ones in **3.5.3** and **4.5.3**. No determinant computation is needed since $\mathrm{Sp}(2m,q) \le \mathrm{SL}(2m,q)$. Since the remainder of the proofs are similar to one another, we will only present one for Proposition 5.18. Once again we arrange to have g normalize Q, but transvections make this easier here and in the next section than it was in the situation of Proposition 4.23.

Define $y \in \{1, j(\gamma)\} \cup \mathcal{S}^*(2)^{j(\gamma)}$ such that $[t^{gy}, t] \ne 1$, as follows: if $[t^g, t] \ne 1$ let $y = 1$; if $[t^g, t^{j(\gamma)}] \ne 1$ let $y = j(\gamma)^{-1}$; if $[t^g, t] = [t^g, t^{j(\gamma)}] = 1$ test the members of $\mathcal{S}^*(2)^{j(\gamma)} \subset Q(\gamma)$ (considering at most one per root group) in order to find one of them, y, such that $[t, t^{gy}] \ne 1$ (i.e., y moves α^g to a point of $\langle \alpha^g, \gamma \rangle$ not in $\alpha^\perp$). Then we have a short straight-line program from $\mathcal{S}^*$ to y (cf. **5.2.1**(iv), **5.2.2**(iv)). Use Lemma 5.12 to find $u \in Q$ such that $Q^{gyu} = Q^{j(\gamma)}$. Use **5.4.3**(2) to obtain uT from $\mathcal{S}^*(2)$ using a straight-line program of length $O(d \log q)$; then observe that short straight-line programs are readily obtained within T. Thus, we have a straight-line program of length $O(d \log q)$ from $\mathcal{S}^*$ to u. Moreover, $gyuj(\gamma)^{-1}$ normalizes T, and if we can find a straight-line program of length $O(d^2 \log q)$ from $\mathcal{S}^*$ to this element then we can find one to g as well.

Now we may assume that g normalizes T and hence also Q. Similarly, use Lemma 5.12 to find $v \in Q$ such that $T(\gamma)^{gv} = T(\gamma)$, use **5.4.3**(2) to obtain vT and then also v from $\mathcal{S}^*(2)$ using a short straight-line program, and replace g by gv.

Now g induces by conjugation an element acting within $\langle T, T(\gamma) \rangle \cong \mathrm{Sp}(2,q)$ and normalizing T and $T(\gamma)$. Use the exponents in our Zech table in **2.3** in order to find the power s^i of the element in **5.2.1**(v) or **5.2.2**(v) such that gs^i commutes with t. As usual we replace g by gs^i.

Use **5.4.3**(4) to find the matrix $\tilde{g}$ representing the linear transformation induced by the action of g on Q/T. Then $\tilde{g} \in \mathrm{Sp}(2m-2,q) = L\lambda_L$ since

g centralizes T. Use Proposition 5.17(b) to find $l := \tilde{g}\lambda_L^{-1}$ by a straight-line program of length $O(m^2 \log q)$. By Lemma 5.16(ii), $\tilde{l} = l\lambda_L = \tilde{g}$, so that $c := gl^{-1}$ centralizes T and $T(\gamma)$, induces 1 on Q/T and hence lies in $\mathrm{Z}(G)$. Thus, we have obtained an element $l \in g\mathrm{Z}(G)$ from $\mathcal{S}^*$. The algorithm ends here in the case of Theorem 1.1. For Theorem 1.1′, use Proposition 5.17(c) to find a straight-line program from $\mathcal{S}^*$ to c. Thus, we have obtained g from $\mathcal{S}^*$, as required.

Time: Dominated by **5.4.3**(4) and Lemma 5.12. □

5.6 Small dimensions and total time

In this section we will complete the proof of Theorems 1.1 and 1.1′ for the groups $\mathrm{PSp}(2m, q)$ and $\mathrm{Sp}(2m, q)$, *assuming* that we know m and q (except for the additional verification using a presentation; cf. **7.2.2**).

5.6.1 $\mathrm{PSp}(4, q)$

We will handle the base case $G \cong \mathrm{PSp}(4, q)$ or $\mathrm{Sp}(4, q)$ *in $O(\xi qe + \mu q \log^2 q)$ time* (cf. **3.2.3**). This is the hardest of the special small-dimensional cases we have toconsider in detail. We may assume that $q \geq 17$.

5.2.1, 5.2.2: Finding a transvection. We will ignore **5.2.2**, instead working towards replacing **5.2.1** for all q.

Lemma 5.19 *There is a Las Vegas algorithm which, with probability $> 1 - 1/2^8$, in $O(\xi q + \mu q \log^2 q)$ time outputs a subgroup T that is $\mathrm{Aut}\,G$–conjugate to a transvection group.*

Note that, when q is even, an outer automorphism of G interchanges transvections and short root elements, which are therefore indistinguishable in this entire algorithm for G: in this case we can obtain a transvection group only up to an outer automorphism.

Proof. Repeat the following procedure up to 2^{17} times:

(I) Choose $\tau \in G$ and test whether it has $\mathrm{ppd}^{\#}(p; e) \cdot \mathrm{ppd}^{\#}(p; 2e)$–order; if q is a Mersenne or Fermat prime we also require that $16\,\|\,|\tau|$. (Then τ induces an element of $\mathrm{ppd}^{\#}(p; e) \cdot \mathrm{ppd}^{\#}(p; 2e)$–order in $G/\mathrm{Z}(G)$.)

Let $a := \tau^{q-1}$.

(II) Choose a conjugate b of a. Use **3.6.2** to test whether $\langle a, b \rangle \cong \Omega^+(4, q)$ or $P\Omega^+(4, q)$ and, if so, then to find the two factors A, $B \cong \mathrm{SL}(2, q)$ or $\mathrm{PSL}(2, q)$ as well as an isomorphism $\lambda_A : A \to \mathrm{SL}(2, q)$ or $\mathrm{PSL}(2, q)$.

Use λ_A to find a Sylow p–subgroup T of A.

We will show that T behaves as required in the lemma.

There are two types of cyclic subgroups of G of the order stated in (I). One type decomposes the underlying vector space V as the perpendicular sum of two nonsingular 2–spaces, while the other decomposes V as the direct sum of two totally singular 2–spaces. View the second type of subgroup within the context of $\Omega(5, q)$: it decomposes the 5–space as the perpendicular sum of two nonsingular 2–spaces and a 1–space; in this case a has 2–dimensional support, and our two conjugates probably generate as stated. On the other hand, if q is odd then two conjugates of the first type can never generate a subgroup $\cong \Omega^+(4, q)$ or $P\Omega^+(4, q)$ [Mi2]. Thus, if the test in (II) succeeds then T is as required.

For future reference we note that A and B act on nonsingular 2–spaces V_A and $V_B = V_A^\perp$ of the target vector space.

Reliability: $\geq 1 - 1/2^8$. For, a single choice τ has the stated order *and the desired type* with probability $\geq (1/2^2)(1/2^2)$ (as in Lemma 2.5), in which case $\langle a, b \rangle$ behaves as desired with probability $\geq 1/640$ by Lemma 4.12(i) (since $q \geq 17$), and then **3.6.2** and **3.6.1** succeed with probability $> 3/4$. Thus, (I) and (II) succeed with probability $> (1/2^4)(1/640)(3/4) > 1/2^{14}$, so at least one of our 2^{17} repetitions succeeds with probability $> 1 - 1/2^8$.

Time: $O(\xi q e + \mu q \log^2 q)$, dominated by **3.6.2** and **3.6.1**. $\square$

Remark. We will find the subgroups in **5.2.1**(ii,iii) in the next lemma, after which we will find the elements in **5.2.1**(iv,v).

5.3.1: Finding $G'_\alpha := \mathrm{C}_G(T)'$ and Q.

Lemma 5.20 *There is a Las Vegas algorithm, running in $O(\xi q e + \mu q \log^2 q)$ time and succeeding with probability $> 1 - 1/2^6$, that finds $\mathrm{C}_G(T)'$ and $Q := O_p(\mathrm{C}_G(T))$.*

Proof. We will use the two subgroups A, B found in (II) above. When (II) was run we obtained a field $\mathbb{F}$ of order q, a generating set S_A^* of A, and an isomorphism $\lambda_A : A \to \mathrm{SL}(2, q)$.

A random conjugate T_0 of T commutes with neither A nor B with probability $1 - 2(q+1)/(q^2+1)(q+1)$. (That is, the center of T_0 is in neither V_A nor V_B, and together with V_B spans a 3–space whose radical is in V_A.)

Let $H := \langle T_0, B \rangle$. Then $Z = \mathrm{Z}(H)$ is a transvection group in A whose center is the aforementioned radical. Moreover, $H = \mathrm{C}_G(Z)'$. For, if 'Q' $= O_p(\mathrm{C}_G(Z)')$, then $\mathrm{C}_G(Z)'$ acts irreducibly on 'Q'$/Z$ by **5.1.3**. Then $H = \mathrm{C}_G(Z)'$ since H contains distinct complements B^{t_0}, $t_0 \in T_0$, to 'Q' in $\mathrm{C}_G(Z)'$. Soon we will find Z and 'Q'.

Test up to 40 elements $\zeta \in H$ in order to find one of $p \cdot \mathrm{ppd}^{\#}(p; 2e)$–order $> 4p$. Since 'Q'$/Z$ is the natural module for $H/$'Q' $\cong \mathrm{SL}(2,q)$ by **5.1.3**, a cyclic subgroup of order $q+1$ acts fixed point freely on 'Q'$/Z$, so that H has at least $|H|/4$ elements of $\mathrm{ppd}^{\#}(p; 2e)$–order > 4 (using Lemma 2.5). Thus, $Z \cap \langle \zeta \rangle \neq 1$, and a single choice for ζ will have the desired order with probability $\geq (1/4)(q-1)/q > 1/5$, so at least one of our choices for ζ has this order with probability $> 1 - 1/2^8$.

We just observed that ζ^{q+1} lies in $\mathrm{Z}(H) < A$. Use λ_A to find $g \in A$ such that $\zeta^{q+1} \in T^g$, and replace T by T^g. Then $H = \mathrm{C}_G(Z)' = \mathrm{C}_G(T)'$.

Let $T_1 := T^{j(\gamma)} \neq T$, where $j(\gamma) \in \mathcal{S}_A^*$ will be used in **5.2.1**(iv).

Let $T_2 := T_0^c$ for $1 \neq c \in T_1$ (so $T_2 \not\leq A$ and T_2 does not commute with T.)

For $i = 1, 2$, let $s_i \in \langle T, T_i \rangle$ have order $q-1$ and normalize T and T_i. These are obtained by using isomorphisms $\langle T, T_i \rangle \to \mathrm{SL}(2,q)$ or $\mathrm{PSL}(2,q)$. We already have such an isomorphism λ_A when $i = 1$, and one for $i = 2$ is obtained with probability $> 1 - 1/2^8$ using up to four applications of **3.6.1**.

Let $Q := \langle T, [s_1, s_2]^{\zeta^{p^k}} \mid 0 \leq k < 2e \rangle$. We claim that $Q =$ 'Q'. By (5.6), s_i acts on 'Q'$/T$ as a scalar. However, s_2 fixes just two nonsingular 2–spaces, hence moves V_A and so does not normalize $T\langle s_1 \rangle$. Then $[s_1, s_2] \notin T$. Since $[s_1, s_2] \in$ 'Q' and ζ^p is irreducible on 'Q'$/T$, it follows from Lemma 2.7 that 'Q'$/T = \langle [s_1, s_2]^{\zeta^{p^k}} T \mid 0 \leq k < 2e \rangle$, and hence that $Q =$ 'Q'.

Since $\mathrm{C}_G(T)$ contains $L := B \cong \mathrm{SL}(2,q)$, we have $\mathrm{C}_G(T)' = QL$ by **5.1.3**.

Reliability: $> 1 - 2/(q^2+1) - 2/2^8 > 1 - 1/2^6$.

Time: $O(\xi qe + \mu q \log^2 q)$, dominated by **3.6.1**. $\square$

As promised above, we can now complete **5.2.1**(iv,v).

(iv) We just defined $j(\gamma)$.

(v) We noted that $s := s_1$ acts on Q/T as a scalar; it centralizes L (which is the group D in **5.2.1**(iii)), and any power of it can be obtained from $\mathcal{S}_A^*$ using a straight-line program of length $O(\log q)$.

5.3.2–5.5: We already found $\lambda_L\colon L \to \mathrm{SL}(2,q)$ or $\mathrm{PSL}(2,q)$ when we called **3.6.2** in (II). Proceed as before, including $\mathcal{S}_A^*$ in $\mathcal{S}^*$.

Reliability: $> 1 - 1/2^6 - 1/2^8 > 1 - 1/2 \cdot 2^2$.

Time: $O(\xi qe + \mu q \log^2 q)$, while **5.5.3** requires times $O(\mu qe)$ and $O(\log q)$ for Theorem 1.1(iii) and (iv), respectively (and likewise for Theorem 1.1$'$(iii,iv)).

As in Corollary 3.20, we have the

Corollary 5.21 *In $O(\xi qe + \mu q \log^2 q)$ time one can test whether or not the group J in **5.2.1** satisfies $J \cong \mathrm{Sp}(4,q)$ and, if so, then find an isomorphism $J \to \mathrm{Sp}(4,q)$ together with the elements and subgroups in **5.2.1**(i-v).*

5.6.2   Total time and reliability

Part (i) of Theorems 1.1 and 1.1$'$ is in Section 7; (ii) is **5.5.2**; (iii) and (iv) are in **5.5.3**; and (vi) is in **5.3.3**. For (vii), by tallying and recalling the recursive call we find that the *total time is*

$$O(m\{\xi(q+m)e \log m + \mu qm^3 \log^2 q \log m\})$$

for (ii), dominated by **5.2.1**, **5.2.2** and **5.3.1**. By **5.5.3**, the straight-line algorithms for (iii) and (iv) take $O(\mu[qe + qm^2] + m^5 \log q)$ and $O(\mu m^2 \log q)$ time, respectively.

Reliability: $> 1/2$.

5.6.3   From $\mathrm{PSp}(4,q)$ to $\Omega(5,q)$

As in **4.6.1**, we now show how to *switch from the 4–dimensional projective representation of $G = \mathrm{PSp}(4,q)$ to the 5–dimensional one of $\Omega(5,q)$*; in fact we will provide two different ways to accomplish this.

Let W be the 4–dimensional module underlying G obtained in **5.6.1**.

Method 1: $\mathrm{Sp}(4,q)$ acts on the orthogonal 6–space $W \wedge W$, and fixes a nonsingular 1–space W_0. Here, W_0 is found using linear algebra as the common eigenspace of all generating elements of $\mathrm{Sp}(4,q)$. The desired $\Omega(5,q)$–module is $W_0^\perp$. **Time:** $O(1)$.

Method 2: We have the associated alternating form on W. Find a standard (hyperbolic) basis of W by solving linear equations. Write down matrices generating the stabilizer $\mathrm{Sp}(4,q)_\Lambda$ of a totally singular line Λ, and also ones generating the subgroup $Q^\$$ of $\mathrm{Sp}(4, q)$ inducing 1 on Λ. Then $Q^\$$ can be viewed as the matrix group "Q" appearing in **4.1.3**. Now proceed exactly as in Section 4.

Outer automorphism: For even q we also can obtain an outer automorphism of $\mathrm{Sp}(4, q)$. Find the radical W_1 of $W_0^\perp$, so $W_0^\perp/W_1$ is a symplectic 4–space. Find standard bases of W and $W_0^\perp/W_1$. Use them to define an isometry $W \to W_0^\perp/W_1$, which induces the desired outer automorphism. **Time:** $O(1)$.

6   Unitary groups: $\mathrm{PSU}(d, q)$

In this section we *change the notation* in Theorems 1.1 and 1.1′: G is $\mathrm{PSU}(V) = \mathrm{PSU}(d, q)$ or $\mathrm{SU}(V)$, where $\mathrm{SU}(V)$ preserves a nonsingular hermitian form (,) on a vector space V of dimension $d \geq 3$ *over* $\mathrm{GF}(q^2)$ *rather than* $\mathrm{GF}(q)$. The involutory field automorphism will be $k \mapsto \bar{k}$. If α is any singular point then $\alpha^\perp / \alpha$ is a unitary space of dimension $d - 2$.

6.1   Properties of G

6.1.1   Transvections

For each singular point $\alpha = \langle u \rangle$, G_α has a normal subgroup consisting of transvections, the (unitary) *transvection group*

$$T(\alpha) = \{v \mapsto v + k(v, u)u \mid \bar{k} = -k\} \cong \mathrm{GF}(q)^+. \tag{6.1}$$

6.1.2   Commutator relations

If α and β are singular points, then

$[T(\alpha), T(\beta)] = 1$ if α and β are perpendicular, and

$\langle T(\alpha), T(\beta) \rangle$ is $\mathrm{SU}(2, q) \cong \mathrm{SL}(2, q)$ if α and β are not perpendicular. In this case $V = \langle \alpha, \beta \rangle \perp \langle \alpha, \beta \rangle^\perp$, and $\langle T(\alpha), T(\beta) \rangle$ induces $\mathrm{SU}(2, q)$ on the first summand and 1 on the second one.

6.1.3 Q

For any singular point α let $Q = Q(\alpha)$ denote the normal subgroup of G_α consisting of all isometries inducing 1 on both α and $\alpha^\perp / \alpha$. If $\alpha = \langle u \rangle$ then Q consists of all maps

$$v \mapsto v + (v, z + lu)u - (v, u)z \quad \text{with } z \in \alpha^\perp, \ l \in \mathrm{GF}(q^2), \ l + \bar{l} + (z, z) = 0.$$

The maps with $z = 0$ comprise $T = T(\alpha) = \mathrm{Z}(Q)$, and $Q/T \cong \alpha^\perp / \alpha$. Alternatively, using a hyperbolic pair u, w (i.e., $(u, u) = (w, w) = 0$, $(u, w) = 1$), Q consists of all maps

$$\begin{aligned} r(z, l) : u \mapsto u, \quad v \mapsto v - (v, z)u \ \forall v \in \langle u, w \rangle^\perp, \quad w \mapsto w + z + lu \\ \text{for } z \in \langle u, w \rangle^\perp \text{ and } l \in \mathrm{GF}(q^2) \text{ satisfying } l + \bar{l} + (z, z) = 0. \end{aligned} \tag{6.2}$$

Once again Q consists of *sparse matrices* with respect to a basis consisting of u, w and a basis of $\langle u, w \rangle^\perp$. We will call elements of conjugates of Q "root elements". A *short root group* is any G-conjugate of $\{ r(kz, kl - \overline{k}\overline{l}) \mid k \in \mathrm{GF}(q^2) \} \cong \mathrm{GF}(q^2)^+$ with $z \neq 0$ and $(z, z) = 0 = l + \overline{l}$.

The following are the same as for the symplectic case:

$$r(z, l)r(z', l') = r(z + z', l + l' - (z, z')) \quad \text{and}$$
$$[r(z, l), r(z', l')] = r(0, (z, z') - (z', z)). \tag{6.3}$$

If ρ is a generator of $\mathrm{GF}(q^2)^*$, define an isometry $s \in \mathrm{GU}(V) - \mathrm{SU}(V)$ by

$$s: u \mapsto \overline{\rho}u, \quad w \mapsto \rho^{-1}w, \quad v \mapsto v \;\; \forall v \in \langle u, w \rangle^\perp, \quad \text{where}$$
$$r(z, l)^s = r(\rho z, \rho\overline{\rho}l) \;\; \text{for } r(z, l) \in Q. \tag{6.4}$$

This turns Q/T into a $\mathrm{GF}(q^2)$-space. If we exclude the case $d = 3$, as well as the cases $d = 4$, $q \leq 3$, then

$$(G_\alpha)' = Q \rtimes (G_{\alpha\gamma})' \text{ where } \gamma = \langle w \rangle, \tag{6.5}$$

Q is *regular on the set of all singular points not in* $\alpha^\perp$, $(G_{\alpha\gamma})'$ *is the special unitary group induced by* $(G_\alpha)'$ *on* $\alpha^\perp/\alpha$, *and the* $(G_{\alpha\gamma})'$*-modules* $\alpha^\perp/\alpha$, $\langle \alpha, \gamma \rangle^\perp$ *and* Q/T *are isomorphic. In fact,*

$$\text{If } g \in (G_{\alpha\gamma})' \text{ then } r(z, l)^g = r(z^g, l) \text{ for all } z \text{ and } l. \tag{6.6}$$

The hermitian form induced on Q/T is given by $(r(z, l)T, r(z', l')T) = (z, z')$. In particular, $(r(z, l)T, r(z, l)T) = 0$ if and only if z is singular.

Lemma 6.7 *Let* $g \in G$ *send* $u \mapsto ju$, $w \mapsto kw$ *(for* $j, k \in \mathrm{GF}(q^2)$*) and induce by conjugation on* Q/T *the linear transformation* $\tilde{g}$. *Then* $\det \tilde{g} = \overline{k}/k^{d-1}$.

Proof. We have $j\overline{k} = 1$ and $w^{g^{-1}r(z,l)g} = w + k^{-1}z^g + l(k\overline{k})^{-1}u$ for any $r(z, l) \in Q$, so that $r(z, l)^g = r(k^{-1}z^g, l(k\overline{k})^{-1})$ where g' denotes the restriction of g to $\langle u, w \rangle^\perp$. It follows that $jk \det g' = 1$ and hence $\det \tilde{g} = (k^{-1})^{d-2} \det g' = \overline{k}/k^{d-1}$. $\square$

6.1.4 Probabilistic generation

Let $d \geq 3$. By [Ka3], every proper irreducible subgroup H of $\mathrm{SU}(d, q)$ generated by transvections is one of the following:

(i) $\mathrm{SU}(d,q')'$ where $q = q'^a$ with a odd;

(ii) $\mathrm{Sp}(d,q')$ where $q = q'^a$;

(iii) $O^\varepsilon(d,q')$ with d and q' even and $q = q'^a$;

(iv) $3\mathrm{P}\Omega^-(6,3)2$ inside $\mathrm{SU}(6,2)$; or

(v) H lies in $\mathbb{Z}_{q+1}\mathrm{wr}S_d$ with $p = 2$.

Only (i) and (ii) can occur if $H = H'$.

The next results are proved as in **3.1.4, 4.1.4, 5.1.4**.

(1) If $d \geq 3$ and $q > 9$ then, with probability $> 1/5$, two nontrivial transvections generate a group J splitting $V = U \oplus U^\perp$ with $\dim U = 2$ and J inducing 1 on $U^\perp$ and inducing on U one of the following: $\mathrm{SU}(2,q)$ and $p \neq 2$, or a dihedral group of order dividing $2(q+1)$ and $p = 2$.

(2) If $d \geq 4$ then, with probability $> 1/2^7$, three nontrivial transvections generate a group J inducing $\mathrm{SU}(3,q)'$ on the nonsingular 3–space $[V,J]$ and 1 on $[V,J]^\perp$.

(3) If $d \geq 4$ then, with probability $> 1/2^9$, four nontrivial transvections generate a group J inducing $\mathrm{SU}(4,q)$ on the nonsingular 4–space $[V,J]$ and 1 on $[V,J]^\perp$.

(4) If $d \geq 6$ then, with probability $> 1/2^{13}$, six nontrivial transvections generate a group J inducing $\mathrm{SU}(6,q)$ on the nonsingular 6–space $[V,J]$ and 1 on $[V,J]^\perp$.

6.1.5 Irreducible and half-irreducible elements

$\mathrm{GU}(2n+1,q)$ contains elements of order $q^{2n+1}+1$, and each is irreducible. As in Lemma 2.5, *at least* $|\mathrm{SU}(2n+1,q)|/4(2n+1)$ *elements of* $\mathrm{SU}(2n+1,q)$ *have* $\mathrm{ppd}^\#(p;2e(2n+1))$–*order*.

$\mathrm{GU}(2n,q)$ contains elements of order $q^{2n}-1$ that split V as $V = U_1 \oplus U_2$ for totally singular n–spaces U_1, U_2 on which the element is irreducible. If n is even then each element of $\mathrm{ppd}^\#(p;2en)$–order fixes two totally singular n–spaces but no nonsingular n–space; as in Lemma 2.5, *there are at least* $|\mathrm{SU}(2n,q)|/4n$ *such elements of* $\mathrm{SU}(2n,q)$. If n is odd then some elements of order $q^n + 1$ decompose V as the sum of two perpendicular nonsingular n–subspaces. In order to prevent the occurrence of such invariant subspaces we will consider elements of $\mathrm{ppd}^{\#}(p; en) \cdot \mathrm{ppd}^{\#}(p; 2en)$–order. As in Lemma 2.5, *there are at least* $|\mathrm{SU}(2n, q)|/8n$ *such elements of* $\mathrm{SU}(2n, q)$.

6.2 Transvections

We now begin our algorithm for Theorems 1.1 and 1.1′ for the unitary groups. The cases $d = 3, 4$ will be handled within our time constraints in **6.6** (cf. **3.2.3**). Thus, for now we *assume that* $d \geq 5$.

6.2.1 Finding transvections

By making up to $\lceil 16qd \ln(4d) \rceil$ choices find $\tau \in G$ of order

$$p \cdot \mathrm{ppd}^{\#}(p; 2e(d-2)) \qquad\qquad \text{if } d \text{ is odd}$$
$$p \cdot \mathrm{ppd}^{\#}(p; e(d-2)) \cdot \mathrm{ppd}^{\#}(p; e(d-2)/2) \quad \text{if } d \text{ is even.}$$

(Recall that this means that $|\tau|$ is divisible by the primes, or 4 or 21, involved in the definition of these $\mathrm{ppd}^{\#}$; cf. **2.4**.) Then τ must split V as $V = V_2 \perp V_{d-2}$ (since, when d is even, the two ppd factors ensure that τ is irreducible on two totally singular $(d-2)/2$–subspaces that span a nonsingular $d-2$–space). If $z := q^{2d-4} - 1$ then $t := \tau^z$ is a transvection.

   In view of **6.1.5**, a single choice of τ has the desired order with probability $\rangle \geq 1/8qd$.

Reliability: $\geq 1 - 1/16d^2$.

Time: $O(qd \log d[\xi + \mu ed^2 \log q])$.

6.2.2 Finding transvection groups and J

As in **5.2.1**, if $q > 2$ make up to $\lceil 2^{11} \ln(4d) \rceil$ choices of three conjugates t_1, t_2, t_3 of t; for each triple use **6.6.2** to test whether $J := \langle t, t_1, t_2, t_3 \rangle \cong \mathrm{SU}(4, q)$ using Corollary 6.19(i). By **6.1.4**(3), with probability $> 1/2^9$ we have $J \cong \mathrm{SU}(4, q)$, in which case the isomorphism test succeeds with probability $> 1/2$; so for each choice we obtain, with probability $> 1/2^{10}$, an isomorphism $\lambda_J : J \to \mathrm{SU}(4, q)$, together with a vector space V_J and generating set $\mathcal{S}_J^*$ of J. Consequently, we obtain these for at least one of our choices with probability $> 1 - 1/16d^2$.

   If $q = 2$ make up to $\lceil 2^{14} \ln(4d) \rceil$ selections of five conjugates of t in order to have t and at least one quintuple generate a subgroup $J \cong \mathrm{SU}(6, 2)$ with probability $> 1 - 1/16d^2$ (cf. **6.1.4**(4)); for each selection, use brute force to test this isomorphism. Then find an isomorphism $\lambda_L \colon J \to \mathrm{SU}(6,2)$, and let $\mathcal{S}_J^*$ denote any set of generators of J consisting of transvections.

Find the following using Corollary 6.19 (or brute force if $q = 2$):

(i) The transvection group T of J containing t, and the center α of $t\lambda_L$;

(ii) $Q_4 := O_p((J\lambda_L)_\alpha)\lambda_L^{-1}$, so that $|Q_4| = q^5$ (unless $q = 2$, in which case $|Q_4| = 2^9$); choose a generating set of Q_4 consisting of short root elements x, chosen so that we also have generators of the root group containing x;

(iii) G_4, the subgroup of $(J\lambda_L)_\alpha\lambda_L^{-1}$ generated by its transvection groups, and a subgroup $D \cong \mathrm{SU}(2,q)$ (or, if $q = 2$, $D \cong \mathrm{SU}(4,2)$) such that $G_4 = Q_4 \rtimes D$ and $D\lambda_J$ fixes a point γ of V_J not perpendicular to α; here $G_4 = G_4'$ if $q \neq 3$;

(iv) $j(\gamma) \in J$, a transvection such that $j(\gamma)\lambda_J$ moves α to γ; and

(v) h, an element of order $q^2 - 1$ that normalizes T and $T(\gamma)$ such that $h\lambda_J$ induces an element of order $q^2 - 1$ on the 1–space α; we can find a straight-line program of length $O(\log q)$ from $\mathcal{S}_J^*$ to any given power of h.

Increasing the set $\mathcal{S}_J^*$: We assume that this is increased to include generating sets of the groups T, Q_4 and G_4 consisting of root elements, as well as the transvection $j(\gamma)$.

Reliability: $\geq 1 - 1/16d^2$.

Time: $O(\log d\{\xi q^2 + \mu q^3 \log q\})$, dominated by **6.6.2**.

6.3   Finding $Q = Q(\alpha)$, $T(\gamma)$, G_α', L and $\mathcal{S}^*$

6.3.1 Q, $T(\gamma)$ and G_α'

Let
$$Q := \langle Q_4^{\tau^{pi}} \mid 0 \leq i \leq d - 2 \rangle$$
$$G_\alpha' := \langle G_4^{\tau^{pi}} \mid 0 \leq i \leq d - 2 \rangle$$
$$Q(\gamma) := Q(\alpha)^{j(\gamma)} \quad \text{and} \quad T(\gamma) := T^{j(\gamma)}.$$

In view of **6.1.3** we may assume that $Q \leq \text{'}Q\text{'}$.

Lemma 6.8   (i) $Q = {}^\lq Q{}^\rq$.

(ii) $G'_\alpha = \mathrm{N}_G(T)' = \mathrm{N}_G(Q)'$.

Proof. We proceed as in the proof of Lemma 5.11, finding that $Q = {}^\lq Q{}^\rq$ and $H := \langle G_4^{\tau^{pi}} \mid 0 \le i < d - 2 \rangle$ induces on the unitary space Q/T an irreducible group generated by transvection groups. Moreover, $G_4 = G'_4$ if $q \ne 3$ (cf. **6.2.2**(iii)), so that the only possibilities for H/Q are $\mathrm{Sp}(d - 2, q)$ and $\mathrm{SU}(d - 2, q)$ in view of **6.1.4**.

If H/Q is $\mathrm{Sp}(d - 2, q)$ then in view of its order, τ^p cannot normalize H, so that $G'_\alpha = \langle H, H^{\tau^p} \rangle > H$. Now the only possibility for G'_α/Q is $\mathrm{SU}(d-2, q)$. $\square$

6.3.2   Effective transitivity of Q; the complement L

Points are defined to be the conjugates $T(\beta)$ of $T = T(\alpha)$. Points $T(\beta)$ and $T(\gamma)$ are called *perpendicular* if and only if $[T(\beta), T(\gamma)] = 1$ (which can be tested by checking just one generator of $T(\beta)$ and one of $T(\gamma)$).

Lemma 6.9 *In deterministic $O(\mu[d + q^3 \log q])$ time an element of Q can be found conjugating any given point T^x not perpendicular to T to any other given point T^y not perpendicular to T.*

Proof. We have $1 \ne t \in T$. Either $[t^x, t^y] \ne 1$ or $[t^x, t^{yt}] \ne 1$. Since $t \in Q$ we may assume that $[t^x, t^y] \ne 1$.

If some $u \in (t^y)^{T^x}$ commutes with t, test one member of each short root group among the generators of Q in order to find a short root group $R < Q$ such that $[R, u] \ne 1$. List the elementary abelian group $[R, u]$, and find one of its q^3 elements v such that $[t^{xv}, t^y] = 1$. Then $v \in Q$ and $T^{xv} = T^y$.

If no $u \in (t^y)^{T^x}$ commutes with t, let $H := \langle T, t^x, t^y \rangle$, so that $H \cong \mathrm{SU}(3, q)'$. Find an isomorphism $\lambda_H \colon H \to \mathrm{SU}(3, q)'$ using Corollary 6.18(ii); use it to find and list $O_p(\mathrm{N}_H(T))$. Find $v \in O_p(\mathrm{N}_H(T))$ such that $[t^{xv}, t^y] = 1$. Then $v \in Q$ and $T^{xv} = T^y$.

Correctness: We have $T^x = T(\gamma)$ and $T^y = T(\delta)$ for singular points $\gamma, \delta \notin \alpha^\perp$. By (6.1), the points α, δ, δ^t are collinear, and there is just one point of their line lying in $\gamma^\perp$. Thus, either $\delta \notin \gamma^\perp$ or $\delta^t \notin \gamma^\perp$; we may assume that $\delta \notin \gamma^\perp$, so that $\langle T^x, T^y \rangle \cong \mathrm{SL}(2, q)$ by **6.1.2**. As usual there is just one element of Q conjugating T^x to T^y; the algorithm finds this element either in $[Q, u]$ or in $O_p(\mathrm{N}_H(T))$.

Let $\varepsilon = \langle \gamma, \delta \rangle \cap \alpha^{\perp}$. This is singular if and only if some $u \in (t^y)^{T^x}$ lies in $T(\varepsilon)$ and hence commutes with t. In this case u induces a transvection of Q/T with center $[Q, u]/T$ (cf. **6.1.3**), and then $[Q, u]$ is elementary abelian of order q^3. A calculation using (6.3) shows that $[R, u] = [Q, u]$, so that listing $[Q, u]$ only takes $O(\mu q^3 e)$ time by Lemma 2.1. Each element of $[Q, u]$ induces a transvection of the 3–space $\langle \alpha, \gamma, \delta \rangle$ having axis $\langle \alpha, \varepsilon \rangle$ (for, u^g is a transvection of V with center on $\langle \alpha, \varepsilon \rangle$ whenever $g \in Q$, and hence $[Q, u]$ fixes all points of $\langle \alpha, \varepsilon \rangle$). Then $[Q, u]$ induces all q^3 transvections of $\langle \alpha, \gamma, \delta \rangle$ with axis $\langle \alpha, \varepsilon \rangle$ that are isometries of $\langle \alpha, \gamma, \delta \rangle$. In particular, some element v of this group sends γ to δ. Here v has center ε and hence fixes the line $\langle \gamma, \delta \rangle$. Since commuting transvections in $\langle T^x, T^y \rangle \cong \mathrm{SL}(2, q)$ lie in the same transvection group, $[t^{xv}, t^y] = 1$ implies that $T^{xv} = T^y$.

If ε is nonsingular then no element of $u \in (t^y)^{T^x}$ commutes with t. Moreover, $\langle \alpha, \gamma, \delta \rangle$ is nonsingular (its radical must lie in $\langle \alpha, \gamma, \delta \rangle \cap \alpha^{\perp} = \langle \alpha, \varepsilon \rangle$, but that line has only one singular point). Consequently, $H = \langle T, t^x, t^y \rangle \cong \mathrm{SU}(3, q)'$. This acts on V by inducing 1 on a nonsingular $d - 3$–subspace, so that $O_p(\mathrm{N}_H(T)) \leq Q$. Here $O_p(\mathrm{N}_H(T))$ has a p–element $v \in Q$ sending γ to δ. Since commuting transvections in H lie in the same transvection group, $[t^{xv}, t^y] = 1$ implies that $T^{xv} = T^y$.

The timing is straightforward except for handling $H \cong \mathrm{SU}(3, q)'$ deterministically. Namely, we already have T as well as $L = \langle T, t^x \rangle \cong \mathrm{SL}(2, q)$. Use Corollary 3.19 to obtain an isomorphism $\lambda_L : L \to \mathrm{SL}(2, q)$ deterministically in $O(\mu q^2 e)$ time. Use this to find λ_H in $O(\mu q^3 \log q)$ time using Corollary 6.18(ii). Use Proposition 6.15 to find $T\lambda_H$ in $O(\mu q^3 e)$ time (only one use of λ_H is needed, since $T\lambda_H$ can be quickly determined from $t\lambda_H$ using linear algebra). List all q^3 matrices in the Sylow p–subgroup of the normalizer of $T\lambda_H$, use λ_H^{-1} to pull all of them back to obtain $O_p(\mathrm{N}_H(T))$ in $O(q^3 \cdot \mu \log q)$ time, and test all $v \in O_p(\mathrm{N}_H(T))$ in $O(\mu q^3)$ time. $\square$

Remark. By (6.5), the element of Q found in the lemma is *unique*, and there is a complement to Q in G'_α. One can be found exactly as in the proof of Corollary 4.18:

Corollary 6.10 *In deterministic $O(\mu[d^2 + q^3 d \log q])$ time a complement $L \cong G'_\alpha/Q \cong \mathrm{SU}(d - 2, q)$ to Q in G'_α can be found centralizing $T(\gamma)$.*

6.3.3   Recursion: $\mathcal{S}^*$ and λ_L

Once again we will require that *our set $\mathcal{S}^*$ of generators of G consists entirely of root elements*. Apply Theorem 1.1′ to L, obtaining *a set $\mathcal{S}_L^*$ of generators*

of L, a vector space V_J over a field $\mathbb{F}$ of size q^2, and a homomorphism $\lambda_L \colon L \to$
SU$(d - 2, q) = $ SU(V_J) *defined on $\mathcal{S}_L^*$. Then $G = \langle G'_\alpha, J \rangle$ is generated by the*
set $\mathcal{S}^* = \mathcal{S}^*(1) \cup \mathcal{S}^*(2) \cup \mathcal{S}^*(3)$ of $O(ed^2)$ elements, where $\mathcal{S}^*(1) = \mathcal{S}_L^*$, $\mathcal{S}^*(2)$ is
*the defining set of $O(ed)$ generators for Q (cf. **6.3.1**) together with generators*
*for T, and $\mathcal{S}^*(3) = \mathcal{S}_J^*$ (cf. **6.2.2**).*

Computing preimages recursively: As in **3.3.3**, we will apply λ_L^{-1} to
elements of $L\lambda_L = $ SU$(d - 2, q)$ in $O(\mu d^2 \log q)$ time using Proposition 6.14
with L in place of G.

The isomorphism $\lambda^\#$:   Once again we are assuming that there exists an
isomorphism $\lambda^\# \colon G \to $ SU(V) or PSU(V), which we may assume sends $Q, Q(\gamma)$
to α, γ in the notation of **6.1.3**. As in **3.4.4** we can use $\lambda^\#$ to see that
λ_L extends to an isomorphism $\lambda^\bullet \colon G'_\alpha = QL \to$ "Q""L" determined by a
semilinear map $Q \to$ "Q"; $\lambda^\bullet|_Q$ also preserves the form on Q/T in the sense
that, for some $k \in \mathbb{F}^*$ and $\sigma \in \text{Aut}\mathbb{F}$, we have $(uT\lambda^\bullet, vT\lambda^\bullet)'' = k(uT, vT)_Q^\sigma$ for
all $u, v \in Q$, where $(\,,\,)''$ and $(\,,\,)_Q$ denote the hermitian forms on "Q"/"T"
and Q/T, respectively. By following $\lambda^\#$ with a field automorphism we see that
we may also assume that $\lambda^\bullet|_Q$ is linear on Q/T; this requirement uniquely
determines $\lambda^\bullet|_Q$ up to a scalar transformation.

6.4   Deterministic algorithms for Q

All procedures in 6.4 and 6.5 will be deterministic.

6.4.1   Managing linear combinations for Q/T

Lemma 6.11 *In deterministic $O(\mu d^2 \log q)$ time $O(e)$–generator subgroups A_i*
$\leq Q$ can be found, where $1 \leq |i| \leq m - 1$ if d is even and $0 \leq |i| \leq m - 1$ if d
is odd, such that the following hold:

(i) *$|A_i T| = q^3$ for each i and Q/T is the direct product of the groups $A_i T/T$;*
and

(ii) *In deterministic $O(\mu d)$ time, any given $uT \in Q/T$ can be expressed in*
the form $uT = \prod_i a_i T$ with $a_i \in A_i$.

Proof. Let $m := \lfloor d/2 \rfloor$. Fix $\kappa \in \mathbb{F}$ such that $\bar{\kappa} = -\kappa \neq 0$. We may assume
that $V_L = \mathbb{F}^{d-2}$ is given as $V_L = W_1 \perp \cdots \perp W_{\lceil (d-2)/2 \rceil}$, where W_i, $1 \leq i \leq$
$m - 1$, is a nonsingular 2–space having a basis e_i, e_{-i} of singular vectors such
that $(e_i, e_{-i}) = \kappa$. Let $c' \in $ SU$(d - 2, q)$ send $e_1 \mapsto \cdots \mapsto e_{m-1} \mapsto e_1$ and

$e_{-1} \mapsto \cdots \mapsto e_{-(m-1)} \mapsto e_{-1}$. Let $j' \in \mathrm{SU}(d-2, q)$ send $e_i \mapsto e_{-i} \mapsto -e_i$ for $1 \le i \le m-1$. We may assume that $e_1, e_{-1}, \ldots, e_{m-1}, e_{-(m-1)}$ are the first $2m-2$ standard basis vectors of $\mathbb{F}^{d-2}$, and that W_m is spanned by the last standard basis vector e_m if d is odd, where $(e_m, e_m) = 1$.

Find the transvections $t'_1 \colon v \mapsto v - \kappa^{-1}(v, e_1)e_1$ and $t'_{-1} := t_1'^{-j'}$ of V_L (cf. (6.1)), having respective matrices $I + E_{-1,1}$ and $I + E_{1,-1}$ in notation corresponding to that in the proof of Lemma 3.13. Find $h'_1 \in \mathrm{SU}(d-2, q)$ inducing $\mathrm{diag}(\rho, \bar{\rho}^{-1}, \rho^{-1}, \bar{\rho}, 1, \ldots, 1)$, unless $d-2 = 3$ and h'_1 induces $\mathrm{diag}(\rho, \bar{\rho}^{-1}, \bar{\rho}/\rho)$ on V_L.

Use Proposition 6.14 to find $c := c'\lambda_L^{-1}$, $j := j'\lambda_L^{-1}$, $t_1 := t'_1\lambda_L^{-1}$ and $h_1 := h'_1\lambda_L^{-1}$. Then find $t_{-1} := t_1^{-j}$, $t_{\pm i} := t_{\pm 1}^{c^{i-1}}$ and $h_i := h_1^{c^{i-1}}$ for $2 \le i \le m-1$.

Find a generator r_1 of Q such that $[r_1, t_1] \ne 1$; let R_1 be the short root group containing r_1 (cf. **6.1.3**, **6.3.1**). Then $[Q, t_1] = [R_1, t_1]$ by (6.3) and hence has $O(e)$ generators. Let

$$
\begin{aligned}
A_1 &:= [[Q, t_1], h_1], \quad A_{-1} := A_1^j \\
A_{\pm i} &:= A_{\pm 1}^{c^{i-1}} \quad \text{for } 2 \le i \le m-1 \\
A_m &:= \langle u[u, t_1]^{-1}[u, t_{-1}]^{-1} \mid u \in \mathcal{S}_Q \rangle \quad \text{if } d \text{ is odd}
\end{aligned}
\tag{6.12}
$$

where $\mathcal{S}_Q$ is our generating set for Q. Then Q/T is the direct product of the $d-2$ groups $A_{\pm i}T/T$ and (if d is odd) A_m/T. Here $|A_m| = q^3$ and $Z(A_m) = T$. Moreover, $A_{\pm i}$ is elementary abelian of order q^2 and $A_{\pm i}^{h_i} = A_{\pm i}$ for $1 \le i \le m-1$ ($A_{\pm i}$ is a short root group): since $h_1 \in L$ centralizes T we have $[Q, t_1] = A_1 \times T$, so that $[A_1, h_1] \le [A_1 T, h_1] \le A_1$.

Given any $uT \in Q/T$ we have

$$
uT = \Pi_1^{m-1}(a_i a_{-i})T \quad \text{or} \quad uT = \{\Pi_1^{m-1}(a_i a_{-i})\}a_m T,
$$

where, whenever $1 \le i \le m-1$,

$$
\begin{aligned}
a_i &:= [[u, t_{-i}], h_i^{-q}]^{j^{-1}h_i^n} \quad \in A_i \\
a_{-i} &:= [[u, t_i], h_i]^{jh_i^{-qn}} \quad \in A_{-i} \\
a_m &:= u\{\Pi_1^{m-1}(a_i a_{-i})\}^{-1} \in A_m \quad \text{if } d \text{ is odd}
\end{aligned}
$$

for an integer n satisfying $\rho^n = (\rho - 1)^{-1}$.

Time: $O(\mu d^2 \log q)$ to find the subgroups A_i and the elements h_i^n and h_i^{-q}, dominated by four uses of Proposition 6.14; and then $O(\mu d)$ to find the components a_i, a_{-i} and a_m.

Correctness: This is essentially the same as in the case $q > 2$ of Lemma 5.14. $\square$

6.4.2   A field of endomorphisms of Q/T

We have $Q = \Pi_1^{m-1}(A_i A_{-i})T$ or $\{\Pi_1^{m-1}(A_i A_{-i})\}A_m$; although these are not direct products, we have uniqueness of the representation of each element of Q as a product in the indicated order (cf. **6.4.1**). When d is odd let ν be an integer such that $(e_1 + \rho^\nu e_{-1}, e_1 + \rho^\nu e_{-1}) = 1$, let $y' \in \mathrm{SU}(W_1 + W_{-1} + W_m)$ with $(e_1 + \rho^\nu e_{-1})^{y'} = e_m$, and find $y := y'\lambda_L^{-1}$. We will define an automorphism $\tilde{s}$ of Q/T of order $q^2 - 1$: if $1 \leq i \leq m - 1$ then

$$\tilde{s} := \begin{cases} h_i & \text{on } A_i T/T \\ h_i^{-q} & \text{on } A_{-i}T/T, \end{cases}$$

where h_i was defined in **6.4.1**. If d is odd then we define $a_m^{\tilde{s}} := a_m^{y^{-1}\tilde{s}y}$ whenever $a_m \in A_m/T$, where $a_m^{y^{-1}}$ lies in $A_m^{y^{-1}} < A_1 A_{-1}T$ (cf. **6.4.1**) and hence its $A_1 T/T$- and $A_{-1}T/T$-components can be found using Lemma 6.11, so that $a_m^{y^{-1}\tilde{s}}$ has already been defined and belongs to $A_1 A_{-1}T/T$.

   In view of the end of **6.3.3**, there is an isomorphism $\lambda^\#\colon Q \to$ "Q" commuting with the actions of L and "L", and this must send some automorphism of Q to an element "s" $\in \mathrm{Aut}(\text{"}Q\text{"})$ inducing scalar multiplication by ρ on "Q"/"T" (namely, $r(z,l) \mapsto r(\rho z, \rho \bar{\rho} l)$; cf. (6.3)) and commuting with the action of "L". The behavior of "s"$\lambda^{\#-1}$ coincides with that of $\tilde{s}$ on each of the subgroups $A_i T/T$, $A_m T/T$, $1 \leq |i| \leq m - 1$. It follows that $\tilde{s}$ commutes with the action of L, and induces an automorphism of Q/T generating the multiplicative group of a field $\mathrm{GF}(q^2)$ of endomorphisms of Q/T. We identify $\tilde{s}$ with the scalar ρ it induces on each of the subgroups $A_i T/T$, A_m/T, $1 \leq |i| \leq m - 1$, thereby identifying the field $\mathbb{F}$ in **2.3** with the present one. While this restriction to $A_1 T/T$ produces our field, the next timing remark is more significant:

Time to apply a field element to $uT \in Q/T$: $O(\mu d)$, using Lemma 6.11(ii) to write uT as a product of $O(d)$ elements of Q/T, on each of which it takes time $O(\mu)$ to apply $\tilde{s}$.

6.4.3 Q/T as an $\mathbb{F}$–space

   (1) *Basis $\mathcal{B}$:* We have an $\mathbb{F}$–basis $e_{\pm 1}, \ldots, e_{\pm(m-1)}$ or $e_{\pm 1}, \ldots, e_{\pm(m-1)}, e_m$ of V_L, where $e_i = e_1^{l'(i)}$ for $1 \leq i \leq m - 1$ for $l'(i) := c'^{i-1}$ or $j'c'^{i-1}$ when $1 \leq i \leq m-1$ and $e_m = (e_1 + \rho^\nu e_{-1})^{y'}$ (cf. **6.4.1**, **6.4.2**). Let $1 \neq b_1 \in A_1 T/T$. Define $\mathcal{B} := \{b_{\pm 1}, \ldots, b_{\pm(m-1)}\}$, or $\{b_{\pm 1}, \ldots, b_{\pm(m-1)}, b_m\}$ if d is odd, where

$b_i := b_1^{l(i)}$ with $l(i) := l'(i)\lambda_L^{-1} = c^{i-1}$ or jc^{i-1} for $1 \le |i| \le m-1$, and $b_m := (b_1 b_{-1}^{\bar{s}^\nu})^y$. Then $\mathcal{B}$ is a basis of Q/T. **Time:** $O(\mu d)$.

Using the action of $\langle \bar{s} \rangle$ we obtain a listing of each A_i/T, as well as a GF(p)–basis $\mathcal{B}_p$ of Q/T. **Time:** $O(\mu q^2 d)$.

(2) *Linear combinations*: Use Lemma 6.11(ii) to express any given $uT \in Q/T$ as an $\mathbb{F}$–linear combination using $\mathcal{B}$, or as a GF(p)–linear combination using $\mathcal{B}_p$, or by means of a straight-line program from $\mathcal{B}_p$ of length $O(d \log q)$. **Time:** $O(\mu q^2 d)$.

(3) *Modify* $\mathcal{S}^*(2)$: We may assume that $\mathcal{S}^*(2)$ contains preimages in $\cup_i A_i$ of all members of $\mathcal{B}_p$.

(4) *Matrices* $\tilde{g}$: Using $\mathcal{B}$ and (2) we can find the matrix $\tilde{g}$ representing the linear transformation induced on the $\mathbb{F}$–space Q/T by any given $g \in \mathrm{N}_G(T)$. **Time:** $O(d \cdot \mu q^2 d)$.

6.4.4   The extension $\lambda^\bullet$

As in **3.4.4**, **4.4.4** and **5.4.4**, we now extend λ_L to G'_α. As in **5.4.4** some care is needed, since Q is not abelian. Let "Q" and "T" denote the matrix groups in **6.1.3**. Recall that, in the notation of (6.2), we have a bijection r from a subset of $\langle \alpha, \gamma \rangle^\perp \times \mathbb{F}$ to "Q" inducing an isometry $\langle \alpha, \gamma \rangle^\perp \to$ "Q"/"T". Let "L" $\cong$ SU$(2m-2, q)$ denote the centralizer of $\langle \alpha, \gamma \rangle$ in the unitary group implicit in **6.1.3**.

We identify V_L with the $d-2$–space $\langle \alpha, \gamma \rangle^\perp$ (in any manner), so that we now have a bijection $r: \{(z,l) \in V_L \times \mathbb{F} \mid l + \bar{l} + (z,z) = 0\} \to$ "Q". In **6.4.1** and **6.4.3**(1) we used the standard basis $e_1, e_{-1}, \ldots$ of V_L and a basis $\mathcal{B} = \{b_1, b_{-1}, \ldots\}$ of Q/T.

Lemma 6.13 *In $O(\mu ed)$ time λ_L can be extended to an isomorphism $\lambda^\bullet: G'_\alpha = QL \to$ "Q" "L" with the following properties:*

(i) *if $b \in Q$ with $bT = \Sigma_i k_i b_i$ (for $k_i \in \mathbb{F}$ found in $O(\mu q^2 d)$ time), then $b\lambda^\bullet = r(\Sigma_i k_i e_i, \rho_b)$ for some $\rho_b \in \mathbb{F}$ that can be found in $O(\mu q)$ time; $\lambda^\bullet: Q/T \to$ "Q"/"T" is $\mathbb{F}$–linear and determines an L–invariant hermitian form $(\,,\,)_{Q/T}$ on Q/T such that $\lambda^\bullet|_{Q/T}$ becomes an isometry; and*

(ii) *$\tilde{l} = l\lambda^\bullet = l\lambda_L$ whenever $l \in L$.*

Proof. At the end of **6.3.3** we saw that there exists such an extension of λ_L inducing a semilinear map on Q/T and uniquely determined up to a scalar transformation. Let $\mathbb{L} := \{l \in \mathbb{F} \mid \bar{l} + l = 0\}$ (cf. (6.2)). We claim that the definition

$$b_{\pm i} \lambda^{\bullet} := r(e_{\pm i}, \mathbb{L}) \quad \text{for } 1 \leq i \leq d - 2$$

$$b_m \lambda^{\bullet} := r(e_m, \mathbb{L}) \quad \text{if } d \text{ is odd}$$

of $\lambda^{\bullet}|_{Q/T}$ is forced by the known actions of L and $L\lambda_L$ (up to the aforementioned scalar ambiguity). For, since $[Q, t_1]\lambda^{\#} = ["Q", t_1\lambda_L] = ["Q", t_1']$ in **6.4.1**, we must have $b_1\lambda^{\bullet} \in r(\langle e_1 \rangle, \mathbb{L})$, and then we may assume that $b_1\lambda^{\bullet} = r(e_1, \mathbb{L})$. Use the elements $l'(i) \in L\lambda_L$ and $l(i) = l'(i)\lambda_L^{-1} \in L$ introduced in **6.4.3**(1) in order to deduce that we must have $b_i\lambda^{\bullet} = (b_1^{l(i)})\lambda^{\bullet} = r(e_1, \mathbb{L})^{l'(i)} = r(e_i, \mathbb{L})$ whenever $1 \leq |i| \leq m - 1$. As in the proof of Lemma 4.20, use the elements h_1 and h_1' in **6.4.1** in order to see that $\lambda^{\bullet}$ is linear on the group generated by all $A_i T/T$, $1 \leq |i| \leq m - 1$. In particular, in the notation of **6.4.3**(1), $(b_1 b_{-1}^{\tilde{s}^\nu})\lambda^{\bullet} = r(e_1 + \rho^\nu e_{-1}, \mathbb{L})$, and hence we must also have $b_m\lambda^{\bullet} = (b_1 b_{-1}^{\tilde{s}^\nu})^y \lambda^{\bullet} = r((e_1 + \rho^\nu e_{-1})^{y'}, \mathbb{L}) = r(e_m, \mathbb{L})$, as claimed.

Now $\lambda^{\bullet}$ is linear on Q/T, and (ii) follows as in **3.4.4**. Moreover, this produces an isomorphism $\lambda^{\bullet}: QL/T \to "Q" "L"/"T"$.

We need three additional ingredients before continuing.

1. *The elements $\hat{b}_i$.* Let

$$\hat{b}_1 \in A_1 \text{ be such that } \hat{b}_1 T = b_1$$

$$\hat{b}_{-1} := \hat{b}_1^j \in A_{-1} \text{ and } \hat{b}_{\pm i} := \hat{b}_{\pm 1}^{c^{i-1}} \in A_{\pm i} \text{ for } 1 \leq i \leq m - 1$$

$$\hat{b}_m := (\hat{b}_1 \hat{b}_{-1}^{\tilde{s}^\nu})^y \text{ if } d \text{ is odd.}$$

Clearly $\hat{b}_i T = b_i$ for all i.

2. *Labeling T.* Start with $t(0) := 1$ and $t(\kappa) := t$. We use the element h_1 in **6.4.2** in order to define $t(\rho^k - \bar{\rho}^k) := [\hat{b}_1^{h_1^k}, \hat{b}_{-1}]$ whenever $0 \leq k < q^2 - 1$. (N.B.—Applying $\lambda^{\bullet}$ will yield $t(\rho^k - \bar{\rho}^k)\lambda^{\bullet} = r(0, \rho^k - \bar{\rho}^k)$.)

Before continuing we digress to determine the action of h_1' on $"Q"$. We know its action on $"Q"/"T"$ (cf. **6.4.1**). By working within the group $"Q" "L"$ we can find an element of $"L"$ sending $r(ke_1, l) \to r(\rho ke_1, l)$ for all k, l, while acting exactly as h_1' does on $"Q"/"T"$ (cf. (6.6)). Hence, h_1' must be this element.

3. *Automorphism s of Q.* We use essentially the same formulas as in **6.4.2**: first let

$$s := \begin{cases} h_i & \text{on } A_i \\ h_i^{-q} & \text{on } A_{-i} \\ t(l) \mapsto t(\rho\bar{\rho}l) & \text{on } T \end{cases}$$

in order to define s on a generating set of $\langle A_i \mid 0 \le i \le m-1 \rangle$; and then also define

$$\hat{b}_m^{s^k} := (\hat{b}_m^{y^{-1}})^{s^k y} = (\hat{b}_1^{s^k} \hat{b}_{-1}^{s^\nu s^k})^y \quad \text{for } 1 \le k \le 2e$$

on a generating set $b_m^{s^k}$, $0 \le k < 2e$, of A_m if d odd.

We claim that this *determines an automorphism s of Q that commutes with the action of L and induces $\tilde{s}$ on Q/T.* For, there is an automorphism $\check{s}$ of G induced by conjugation by the preimage "s"$\lambda^{\#-1}$ of the element in (6.4); $\check{s}$ normalizes $T(\gamma)$ and commutes with L (since $\lambda^\#$ sends α, γ to $T(\alpha), T(\gamma)$). By **6.4.2**, $\check{s}$ and $\tilde{s}$ both induce scalar multiplication by ρ on Q/T. We just saw that $r(\langle e_1 \rangle, 0)$ is h_1'-invariant. Then $r(\langle e_1 \rangle, 0)\lambda^{\#-1}$ must be h_1-invariant, and hence is A_1. Similarly, each $r(\langle e_{\pm i} \rangle, 0)\lambda^{\#-1}$ must be h_i-invariant for $1 \le |i| \le m-1$.

Let $1 \le i \le m-1$. Since h_i and $\check{s}$ agree on $A_i T/T$ it follows that they agree on A_i. Similarly, s and $\check{s}$ agree on A_{-i}. Finally, since $\check{s}$ must commute with y we have $\hat{b}_m^{s^k} := (\hat{b}_m^{y^{-1}})^{\check{s}^k y} = (\hat{b}_1^{\check{s}^k} \hat{b}_{-1}^{s^\nu \check{s}^k})^y = (\hat{b}_1^{s^k} \hat{b}_{-1}^{s^\nu s^k})^y$ for $1 \le k \le 2e$. Thus, s and $\check{s}$ coincide on a generating set of Q, as claimed. (Soon we will make the definition of s effective on all of Q.)

Moreover, we just saw that $\lambda^\#$ extends λ_L to an isomorphism $QL\langle s \rangle \to$ "Q" "L"$\langle$"s"$\rangle$ such that $s\lambda^\# =$ "s"; of course, this extension will not be effective until we have completed the proof of the lemma.

Now we can continue the proof of the lemma. Define $\lambda^\bullet$ on Q as follows: whenever $0 \le k < 2e$, $1 \le |i| \le m-1$ and $l \in \mathbb{L}$,

$$\hat{b}_i^{s^k} \lambda^\bullet := r(\rho^k e_i, 0)$$
$$\hat{b}_m^{s^k} \lambda^\bullet := r(\rho^k e_m, \rho^k \bar{\rho}^k \rho^\nu \kappa) \quad \text{if } d \text{ is odd}$$
$$t(l) \lambda^\bullet := r(0, l).$$

Note that these definitions are essentially forced by the existence of $\lambda^\#$: we just saw that $A_i \lambda^\bullet = r(\langle e_i \rangle, 0)$; since $\hat{b}_i \lambda^\bullet$ projects onto e_i it follows that $\hat{b}_i \lambda^\bullet = r(e_i, 0)$, and then also $\hat{b}_i^{s^k} \lambda^\bullet = (\hat{b}_i \lambda^\#)^{(s\lambda^\#)^k} = r(e_i, 0)^{"s"^k} = r(\rho^k e_i, 0)$ using the known extension of λ_L to $QL\langle s \rangle$ together with the definition of "s". Similarly, $\hat{b}_m^{s^k} \lambda^\bullet = (\hat{b}_1^{s^k} \hat{b}_{-1}^{s^\nu s^k})^y \lambda^\bullet = \{r(\rho^k e_1, 0) r(\rho^k \rho^\nu e_{-1}, 0)\}^{y'} = r(\{\rho^k e_1 + \rho^k \rho^\nu e_{-1}\}^{y'}, -\rho^k \rho^\nu \bar{\rho}^k \bar{\kappa}) = r(\rho^k e_m, \rho^k \bar{\rho}^k \rho^\nu \kappa)$ by (6.3).

Thus, we have reconstructed the isomorphism $\lambda^\#|_{QL}$: it is, indeed, $\lambda^\bullet$ as defined above. Since we know that the forms on Q/T and "Q"/"T" must be preserved up to a scalar and a field automorphism, while $\lambda^\bullet$ is linear on Q/T, we can use $\lambda^\bullet$ to determine the desired form on Q/T.

Finally, our definition of $\lambda^\bullet$ on $G'_\alpha = QL$ is *effective on Q*. Namely, let $g \in Q$. Use **6.4.3**(2) to write $gT = xT$ for a product x of powers $\hat{b}_i^{s^k}$, $0 \le k < 2e$. Find gx^{-1}, and write it as $t(l) \in T$ (recall that we have listed T). Then $g\lambda^\bullet = r(0, l)(x\lambda^\bullet)$, as required for (i).

Time: For the definition of $\lambda^\bullet$ this is dominated by the time $O(\mu e d + e d \cdot d^2)$ to write all of the matrices $\hat{b}_i^{s^k} \lambda^\bullet$, as well as $t(l)$ for a GF(p)–basis of $\mathbb{L}$; finding $g\lambda^\bullet$ uses **6.4.3**(2). $\square$

Remark. The spaces Q/T and V_L are related via $u = \sum_i k_i b_i \mapsto \dot{u} = \sum_i k_i e_i$, so that

$$b\lambda^\bullet = r(\dot{bT}, \rho_b) \quad \text{for all } b \in Q \text{ and some } \rho_b \in \mathbb{F}.$$

6.5   The homomorphism λ and straight-line programs

6.5.1   Labeling points

Labeling any given point by a 1–space in $V := \mathbb{F}^d$ in a G'_α–invariant manner is achieved exactly as in **5.5.1**.

Time: $O(\mu[q^3 \log q + q^2 d])$ to label one point, dominated by Lemmas 6.9 and 6.13(i).

6.5.2   The homomorphism and the form

A homomorphism λ from G onto $\mathrm{PSU}(V)$ also is obtained as before. **Time:** $O(ed\mu[q^3 \log q + q^2 d] + ed^2 \cdot d^3 \log q)$ *to find the set $\mathcal{S}^* \lambda$ of elements of order p in addition to the time to find $\mathcal{S}_L^*$*. Equation (4.21) still holds, and $\lambda = \lambda^\bullet$ on Q. A hermitian form on V is found exactly as in **4.5.2**.

6.5.3   Theorems 1.1(iii,iv) and 1.1$'$(iii,iv)

Proposition 6.14 *In deterministic $O(d^4 \log q)$ time, given $h \in \mathrm{GL}(d, q^2)$,*

(a) *it can be determined whether or not $h \in G\lambda$ (modulo scalars), and*

(b) *if so then a straight-line program of length $O(d^2 \log q)$ can be found from $\mathcal{S}^* \lambda$ to h (only modulo scalars in the situation of Theorem 1.1(iv)), after which $h\lambda^{-1}$ can be found in additional $O(\mu d^2 \log q)$ time.*

Moreover,

(c) *in Theorem 1.1$'$, $\mathrm{Z}(G)$ can be found in $O(\mu d^2 \log q)$ time.*

Proposition 6.15 *In deterministic $O(\mu[q^3 \log q + q^2 d^2])$ time, a straight-line program of length $O(d^2 \log q)$ can be found from $\mathcal{S}^*$ to any given $g \in G$. (Hence, the corresponding straight-line program can be written from $\mathcal{S}^*\lambda$ to $g\lambda$, after which $g\lambda$ can be found in additional $O(d^3 \cdot d^2 \log q)$ time.)*

Since the proofs are similar to those for **5.5.3**, we will only present a proof of the second of these propositions.

Exactly as in the proof of Lemma 5.18 we can reduce to the case in which g normalizes both T and $T(\gamma)$.

Let h be the element of J in **6.2.2**(v). Use **6.4.3**(4) to find matrices $\tilde{g}$ and $\tilde{h}$ representing the elements of $\mathrm{GU}(d-2, q)$ induced by the actions of g and h on Q/T. By Lemma 6.7, $\det \tilde{h}$ generates the determinants of the transformations induced on Q/T by all of the elements of $\mathrm{N}_G(T)$. Use the exponents in our Zech table in **2.3** to find a power h^i such that $\det(\tilde{g}\tilde{h}^i) = 1$, and use **6.2.2**(v) to find a straight-line program from $\mathcal{S}^*$ to h^i. Replace g by gh^i in order to have $\tilde{g} \in \mathrm{SU}(d-2, q) = L\lambda_L$.

Use Proposition 6.14(b) to find $l := \tilde{g}\lambda_L^{-1}$ by a straight-line program of length $O(d^2 \log q)$ from $\mathcal{S}_L^*$. By Lemma 6.13(ii), $\tilde{l} = l\lambda_L = \tilde{g}$, so that $c := gl^{-1}$ induces 1 on Q/T, normalizes T and $T(\gamma)$, and hence lies in $\mathrm{Z}(G)$. Thus, we have obtained an element $l \in g\mathrm{Z}(G)$ from $\mathcal{S}^*$. The algorithm ends here in the case of Theorem 1.1. For Theorem 1.1', use Proposition 6.14(c) to find a short straight-line program from $\mathcal{S}^*$ to c. Thus, we have obtained g by a straight-line program from $\mathcal{S}^*$, as required.

Time: Dominated by Lemma 6.9 and **6.4.3**(4). $\square$

6.6   Small dimensions and total time

In this section we will complete the proof of Theorems 1.1 and 1.1' for the unitary groups $\mathrm{PSU}(d, q)$ and $\mathrm{SU}(d, q)$, *assuming* that we know d and q (except for the additional verification using a presentation; cf. **7.2.2**).

6.6.1   PSU$(3, q)$

We now handle the base case $G \cong \mathrm{PSU}(3, q)$ or $\mathrm{SU}(3, q)$ in $O(\xi q^2 + \mu q^3 \log q)$ time. We may assume that $q > 5$. This is similar to, but harder than, **3.6.1**; it is the most difficult part of Section 6.

We begin by finding a subgroup $L \cong \mathrm{SL}(2, q)$ using the method in **3.6.3**. Test up to 128 elements τ in order to find one of $\mathrm{ppd}^{\#}(p; 2e) \cdot \mathrm{ppd}^{\#}(p; e)$–

or $\mathrm{ppd}^{\#}(p; 2e) \cdot \mathrm{ppd}^{\#}(p; e) \cdot \mathrm{ppd}^{\#}(p; e/2)$–order, respectively, depending on whether e is odd or even; in addition we require that $|\tau^{2(q-1)}| > 3$, and if q is a Mersenne or Fermat prime also that $16\||\tau$. (An element has this order with probability $> 1/16$, so we will succeed with probability $> 1 - 1/2^8$.)

Let $a := \tau^{2(q-1)}$. (Then a will have an i–dimensional eigenspace V_i for $i = 1, 2$, where $V_2 = V_1^{\perp}$. If $g \in G$ and $V_1 \neq V_1^g \not\subseteq V_2$, then $\langle a, a^g \rangle$ preserves the decomposition $(V_2 \cap V_2^g) \perp \langle V_1, V_1^g \rangle$ and acts irreducibly on $\langle V_1, V_1^g \rangle$. With high probability, $\langle a, a^g \rangle$ induces at least $\mathrm{SU}(2, q) = \mathrm{SL}(2, q)$ on this subspace as in Lemma 3.8(i).)

Choose up to 2 conjugates b of a. With probability $1 - (q^2 - q + 1)/(q^2(q^2 - q + 1) > 0.05$, for a single b we will have $A := \langle a, b \rangle \cong \mathrm{SL}(2, q) \circ \langle a \rangle$, and hence this will hold with probability $> 1 - 1/2^8$ for at least one of our choices. Find the subgroup $L \cong \mathrm{SL}(2, q)$ of A as in **3.6.3**, as well as a field $\mathbb{F}$ of size q and an isomorphism $\lambda_L \colon L \to \mathrm{SL}(2, q)$ and a generating set $\mathcal{S}_L^*$ of L behaving as in **3.6.1**.

List L as follows: first list all $O(q^3)$ matrices in $L\lambda_L$; then apply λ_L^{-1} to each of them. Find the Sylow p–subgroup T of L containing t.

Find a conjugate j of t lying in $L^{\tau} - T$ (while a normalizes L, τ does not). Then $T_0 := T^j$ is not in L. We obtain the set T_0^L of $q^3 - q$ subgroups as follows: fix a set $\mathcal{S}_0$ of e generators of T_0, and then T_0^L is $\{\langle \mathcal{S}_0^l \rangle \mid l \in L\}$. (Note that different elements l produce different subgroups $\langle \mathcal{S}_0^l \rangle$.)

Although G acts 2–transitively on the set $\Omega := T^L \cup T_0^L$ of size $q^3 + 1$, we do not know how to determine the permutation action of the generators of G in a direct fashion. Instead, first we will construct the 3–dimensional (projective) matrix representation for G using the fact that each member of Ω contains exactly one member of $Y := \{t, t^{jl} \mid l \in L\}$. Since $l \mapsto t^{jl}$, $l \in L$, is bijective, our listing of L yields one of Y.

Find a field $\mathbb{F}^{(2)}$ of size q^2 as a 2–dimensional vector space over $\mathbb{F}$; operations in $\mathbb{F}^{(2)}$ take $O(1)$ time. Now extend λ_L to a monomorphism $\lambda_L \colon L \to \mathrm{SL}(3, q^2)$ such that $L\lambda_L$ preserves a nonsingular hermitian form on $V := (\mathbb{F}^{(2)})^3$ and preserves the decomposition $V = V_2 \perp V_1$, where V_2 consists of the vectors with last coordinate 0 and $V_1 = \langle 0, 0, 1 \rangle$.

Let $T_1 := T$. Take any conjugate $T_2 \neq T_1$ of T_1 inside L. Use λ_L to label the $q + 1$ conjugates in T^L using 1–spaces of V_2. We may assume that T_1 and T_2 are labeled $\langle v_1 \rangle = \langle 1, 0, 0 \rangle$ and $\langle v_2 \rangle = \langle 0, 1, 0 \rangle$, respectively.

Take any singular 1–space $\langle v_0 \rangle$ of $V - V_2$ and let it be the label of T_0. (Since L is transitive on the singular points of $V - V_2$ this choice can arbitrary.)

For any given $l \in L$ we label T_0^l by $\langle v_0 \rangle (l\lambda_L)$. Pick any one of the $q^3 + 1$

members T_3 of Ω not in $L \cup T_0^{T_1} \cup T_0^{T_2}$. (Then T_0, T_1, T_2, T_3 are in "general position".)

For any given $g \in G$ we label T^g as follows: if t^g is in L then we have labeled T^g already; otherwise T^g is labeled $\langle v_0 \rangle (l\lambda_L)$ where $l \in L$ and $[t^g, t^{jl}] = 1$. Labeling T^g takes $O(\mu q^3)$ time. In particular, find the label $\langle v_3 \rangle$ of T_3.

For any given $x \in G$ we obtain a matrix $\widehat{x}$ in $O(\mu q^3)$ time as follows. For $i = 0, 1, 2, 3$, find the label $\langle w_i \rangle$ of T_i^x, so that $\widehat{x}$ sends $\langle v_i \rangle$ to $\langle w_i \rangle$. As in **3.5.2** this uniquely determines the matrix $\widehat{x}$ up to scalars. Now define $x\lambda := \widehat{x}$ modulo scalars.

Let $\mathcal{S}^*$ consist of the $O(e)$ generators of L obtained in **3.6.1** (using generators of T and T_2) together with the additional transvection j. Find $\mathcal{S}^*\lambda$.

Reliability: $> 1 - 4/16 \cdot 3^2$.

Time: $O(\xi q^2 + \mu q^3 \log q)$, dominated by finding and then listing L by applying λ_L^{-1} to $O(q^3)$ matrices.

Remark. We are indebted to R. A. Liebler for a suggestion that led to this fast timing.

Images of elements of order p. If $|x| = p$, replace $x\lambda$ by $(x\lambda)^{1-q}$ in order to obtain a matrix of order p. In particular, we now have $\mathcal{S}^*\lambda$ and $T\lambda$ without any scalar ambiguities, and $G\lambda$ is $\mathrm{PSU}(3, q)$ or $\mathrm{SU}(3, q)$. Moreover, for any $x \in G$ for which we have a straight-line program from $\mathcal{S}^*$, we can now obtain an unambiguous image $x\lambda$. Similarly, if $y \in \mathrm{SU}(3, q)$ and we have a straight-line program from $\mathcal{S}^*\lambda$ to y, we can obtain a preimage $y\lambda^{-1}$.

Sylow p–subgroups of the normalizers of $T\lambda$ and T. We need to find preimages of elements of $\mathrm{PSU}(3, q)$ or $\mathrm{SU}(3, q)$, as well as the straight-line programs in Theorems 1.1 and 1.1′. However, first we construct needed ingredients related to the Sylow p–subgroups $Q\lambda$ and Q of the normalizers of $T\lambda$ and T, respectively. It is easy to write down matrices generating $Q\lambda$, but we need to obtain these matrices from our set $\mathcal{S}^*\lambda$ using short straight-line programs. Recall that $\mathcal{S}^*$ contains generators for T_1 and T_2, and that generators for $T_0 = T_1^j$ have length 3 as words in $\mathcal{S}^*$.

The three matrix groups $T_i\lambda$, $i = 0, 1, 2$, produce three groups $L_0 := L\lambda = \langle T_1\lambda, T_2\lambda \rangle$ and $L_i := \langle T_0\lambda, T_i\lambda \rangle$ for $i = 1, 2$, all isomorphic to $\mathrm{SL}(2, q)$. In the matrix group L_i, $i = 1, 2$, find a straight-line program of length $O(\log q)$ from $L_i \cap \mathcal{S}^*\lambda$ to a generator h_i of $\mathrm{N}_{L_i}(T_0\lambda) \cap \mathrm{N}_{L_i}(T_i\lambda)$ (use Corollary 3.19 with $\mu = O(1)$; alternatively, this involves straightforward calculations with 3×3 matrices in $G\lambda$).

Then generators of $T_0\lambda$ together with the elements $[h_1, h_2]^{h_1^i}$, $0 \leq i < e$, generate an abelian subgroup Q_{12} of order q^2 in $\mathrm{N}_{G\lambda}(T_0\lambda)$, by Lemma 2.7. (Namely, $\langle h_1, h_2 \rangle$ lies inside the normal subgroup of $\mathrm{N}_{G\lambda}(T_0\lambda)$ of order $q^3(q-1)$, where h_i acts as scalar multiplication on $Q_0/\mathrm{Z}(Q_0) \cong \mathbb{F}^2$ for the Sylow p–subgroup Q_0 of $\mathrm{N}_{G\lambda}(T_0\lambda)$.)

The Q_{12}–conjugates of $T\lambda$ correspond to q^2 singular points of V lying on q lines through the center $\langle v_0 \rangle$ of $T_0\lambda$. Find $u \in T\lambda$ such that $U := (T_2\lambda)^u$ corresponds to one of the $q+1$ singular points of $V_2 = [V, L\lambda]$ *not* among these q^2 points; we may assume that $u \in \mathcal{S}^*\lambda$. As above, find a generator h_3 of $\mathrm{N}_{\langle T_0\lambda, U \rangle}(T\lambda) \cap \mathrm{N}_{\langle T_0\lambda, U \rangle}(U)$ by means of a straight-line program from $\mathcal{S}^*\lambda$, and find a subgroup Q_{13} of $\langle h_1, h_3 \rangle$ of order q^2.

Let $X := Q_{12}Q_{13}$. This has order q^3 since $Q_{12}/T_0\lambda$ and $Q_{13}/T_0\lambda$ are distinct 1–spaces of the 2–dimensional $\mathbb{F}$–space $Q_0/T_0\lambda$ (cf. **6.1.3**). We have generators for these 1–spaces as $\mathrm{GF}(p)$–spaces, and likewise for $T_0\lambda$. Thus, in $O(\mu \log q)$ time a straight-line program of length $O(\log q)$ can be found from $\mathcal{S}^*\lambda$ to any given element of Q_{12} or Q_{13}, and hence to any given element of X since $X/T_0\lambda = Q_{12}/T_0\lambda \times Q_{13}/T_0\lambda$.

Consequently, by mirroring these straight-line programs in G we see that the same statements hold for the Sylow p–subgroup $Q := (X\lambda^{-1})^{j^{-1}}$ of $\mathrm{N}_G(T)$, as well as for $Q\lambda = X^{j^{-1}\lambda} \lhd \mathrm{N}_{G\lambda}(T\lambda)$.

The normalizers of $T\lambda$ and T. We still need to find elements of order $q^2 - 1$ or $(q^2 - 1)/(3, q + 1)$ in these normalizers using straight-line programs from $\mathcal{S}^*\lambda$ or $\mathcal{S}^*$. We write matrices for $Q\lambda$ using the ordered basis $(1, 0, 0)$, $(0, 0, 1)$, $(0, 1, 0)$ of V. (Recall that T_1 and T_2 correspond to the first and last of these vectors.) Consider the matrices

$$x(c, d) := \begin{pmatrix} 1 & 0 & 0 \\ c & 1 & 0 \\ d & -\bar{c} & 1 \end{pmatrix}, h(k) := \begin{pmatrix} k & 0 & 0 \\ 0 & \bar{k}/k & 0 \\ 0 & 0 & 1/\bar{k} \end{pmatrix}, r := \begin{pmatrix} 0 & 0 & 1 \\ 0 & -1 & 0 \\ 1 & 0 & 0 \end{pmatrix} \quad (6.16)$$

with $d + \bar{d} + c\bar{c} = 0$ and $k \neq 0$. Then $Q\lambda$ consists of the matrices $x(c, d)$ (cf. **6.1.3**); also, $(T\lambda)^r = T_2\lambda$ and $r \in L_0$, so that r can be obtained by a straight-line program from $\mathcal{S}^*\lambda$ in $O(\log q)$ time. Let d be a generator of $(\mathbb{F}^{(2)})^*$, find $c \in \mathbb{F}^{(2)}$ satisfying $d + \bar{d} + c\bar{c} = 0$, and note that

$$h(d) = x(c/d, 1/\bar{d})r\, x(c, d)\, rx(c/\bar{d}, 1/\bar{d})r \quad (6.17)$$

generates $\mathrm{N}_{G\lambda}(T_1\lambda) \cap \mathrm{N}_{G\lambda}(T_2\lambda)$, while $\mathrm{N}_{G\lambda}(T_1\lambda) = Q\lambda \rtimes \langle h(d) \rangle$. It follows that, in $O(\log q)$ time, we can obtain $h(d)$ by a straight-line program from $\mathcal{S}^*\lambda$; and hence we can also obtain $h := h(d)\lambda^{-1}$ by a straight-line program from $\mathcal{S}^*$. In particular, we also obtain both $\langle h \rangle = \mathrm{N}_G(T_1) \cap \mathrm{N}_G(T_2)$ and $\mathrm{N}_G(T_1) = Q\langle h \rangle$ in $O(\mu \log q)$ time.

Theorems 1.1(iii,iv) and 1.1'(iii,iv)   (i.e., Propositions 6.15 and 6.14): Once again these are similar, and hence we will only present (iii). We are given $g \in G$. Find $x \in L$ such that $t^x \in t^L \cup t^{jL}$ commutes with t^g, and find a straight-line program of length $O(\log q)$ from $\mathcal{S}_L^*$ to x. Then gx^{-1} normalizes T, and we can replace g by gx^{-1}.

Now $T^g = T$. Find $y \in Q$ such that $[T_2^{gy}, T_2] = 1$ (recall that Q is transitive on $\Omega - \{T\}$). Then $gy \in \mathrm{N}_G(T) \cap \mathrm{N}_G(T_2) = \langle h \rangle$. Finally, any element of $\langle h \rangle$, and in particular gy, can be obtained by a straight-line program of length $O(\log q)$ from $\mathcal{S}^*$ in $O(\mu q^2)$ time.

Time: $O(\mu \log q)$ and $O(\mu q^3 e)$ for Propositions 6.14 and 6.15, respectively.

Corollary 6.18   (i) *In Las Vegas $O(\xi q^2 + \mu q^3 \log q)$ time one can test whether or not the group J in 6.6.2 satisfies $J \cong \mathrm{SU}(3,q)$, and, if so, then obtain the following: an isomorphism λ_J and generating set $\mathcal{S}_J^*$ required there; and, in $O(\mu q^3 e)$ time, a list of $e(q^3 + 1)$ transvections in J that contains e generators of any given transvection group in J.*

  (ii) *In deterministic $O(\mu q^3 \log q)$ time, given a black box group $G \cong \mathrm{SU}(3,q)$ having $O(e)$ generators, and also a subgroup $L \cong \mathrm{SL}(2,q)$ of G generated by transvection groups together with an isomorphism $\lambda_L \colon L \to \mathrm{SL}(2,q)$ behaving as in 3.6.1, one can find an isomorphism $\lambda \colon G \to \mathrm{SU}(3,q)$ behaving as in Theorem 1.1'.*

Proof. (i) Most of this is proved as in Corollary 3.20. In order to list transvections, we use the notation $L \cong \mathrm{SL}(2,q)$ and T_0 in the algorithm in this section. We need to list the $e(q^3 + 1)$ nontrivial elements in the union of $T^L \cup T_0^L$. We already have listed L and hence its transvections, so that the required $e(q+1)$ transvections in L are easily found. For the remaining transvections, list a set A_0 of e generators of T_0 and, whenever $1 \neq t_0 \in A_0$, list $\{t_0^l \mid l \in L\}$; once again $l \mapsto t_0^l$ is bijective, and these lists are pairwise disjoint. Now concatenate the lists.

  (ii) Probability entered into the algorithm earlier in this section four times: finding a transvection t; finding a subgroup $L \cong \mathrm{SL}(2,q)$ generated by transvection groups; finding an isomorphism $\lambda_L \colon L \to \mathrm{SL}(2,q)$ behaving as in 3.6.1; and finding a conjugate $t^j \notin L$. The first three of these are part of the present hypotheses, and allow us to list all transvections in L by using λ_L^{-1} to pull back all transvections of $\mathrm{SL}(2,q)$. In order to find t^j, test each generator j of G to see if t^j is on the aforementioned list of transvections of L. Now proceed exactly as before. $\square$

6.6.2 PSU(4, q)

Finally, we handle the base case $G \cong \mathrm{PSU}(4,q)$ or $\mathrm{SU}(4,q)$ in $O(\xi q^2 + \mu q^3 e \log q)$ time. We may assume that $q \neq 3$.

6.2.1–6.3.1: As in **6.2.1**, choose up to $96q$ elements to find one of $p \cdot \mathrm{ppd}^\#(p;2e) \cdot \mathrm{ppd}^\#(p;e)$–order (this means that $|\tau|$ is divisible by the integers involved in the definition of these $\mathrm{ppd}^\#$; cf. **2.4**); we also require that $\tau^{8p(q-1)} \neq 1$ if q is a Mersenne prime. As in **3.2.1** we find that $t := \tau^z$ is a transvection, where $z := q^2 - 1$. (An element has this order with probability $> 1/16$, so we will succeed with probability $> 1 - 1/2^8$.)

Select 2^{11} pairs t_1, t_2 of conjugates of t, and for each use Corollary 6.18(i) to test whether $J := \langle t, t_1, t_2 \rangle \cong \mathrm{SU}(3,q)$. By **6.1.4**(2), this isomorphism holds for a single pair with probability $> 1/2^7$, in which case the isomorphism test then succeeds with probability $> 1/2$. Thus, with probability $> 1 - 1/(2.7)^{2^3} > 1 - 1/2^{11}$, for at least one pair we obtain an isomorphism $\lambda_J \colon J \to \mathrm{SU}(3,q)$, a field $\mathbb{F}$ of order q^2, and a list consisting of a set X of $O(q^3 e)$ transvections of J such that each transvection group of J is generated by a subset of X. Store X so as to keep track of which elements lie in the same transvection group.

Find $t\lambda_J$. Use $O(e)$ applications of λ_J^{-1} to 3×3 matrices in $J\lambda_J$ in order to find the following: the transvection group T containing t as well as $\mathrm{N}_J(T)$ and its Sylow p–subgroup Q_3. We still need to find the larger group 'Q' $:= O_p(\mathrm{N}_G(T)) > Q_3$ of order q^5 (cf. **6.1.3**).

We have $X^\tau \neq X$. Replace t by a conjugate in X in order to have $t \notin X^\tau$. It follows that $L := \langle \mathrm{C}_{X^\tau}(t) \rangle \cong \mathrm{SL}(2,q)$. (For, J^τ corresponds to a 3–space $W = [V, J^\tau]$ of the target vector space V. Any singular point not in W is perpendicular to a nonsingular line of W. If a transvection centralizes t then so does its entire transvection group, for which X^τ contains generators.)

Let $y \in \mathrm{C}_{X^\tau}(t)$ and $Q := \langle Q_3, Q_3^y \rangle$. Then $Q =$ 'Q'. (For, since $Q_3 \leq$ 'Q' also $Q \leq$ 'Q'. Since Q_3/T is a nonsingular 1–space of the 2–dimensional $\mathbb{F}$–space 'Q'$/T$ (cf. **6.1.3**) it is moved by y and hence $Q/T =$ 'Q'$/T$.)

Let $G'_\alpha := \langle Q, L \rangle = Q \rtimes L$; this is $\mathrm{N}_G(T)'$ by **6.1.3**.

We need to consider **6.2.2**(i-v). We have already taken care of (i), while (ii) and (iii) are to be omitted except for the definition $D := L$. For (v), find

$h \in J \cong \mathrm{SU}(3, q)$. For (iv), use any transvection $j(\gamma) \in J - T$, so that T and $T(\gamma) := T^{j(\gamma)}$ are not perpendicular.

6.3.2–6.5: Proceed as before. The recursive call in **6.3.3** is to **3.6.1**.

Reliability: $\geq 1 - 1/2^8 - 1/2^{11} - 1/8^3 > 1 - 1/4 \cdot 4^2$, including up to three calls to **3.6.1** in order to identify a new group L constructed in **6.3.2**. (A new L is needed so as to avoid having to find a second transvection group centralized by the old L.)

Time: $O(\xi q^2 + \mu q^3 e \log q)$ for Theorem 1.1(ii), dominated by Corollary 6.18 and finding $\mathcal{S}^* \lambda$. By Propositions 6.15 and 6.14, the timing is $O(\mu q^3 e)$ and $O(\mu \log q)$ for Theorem 1.1(iii) and (iv), respectively (as well as for Theorem 1.1'(iii,iv)).

Corollary 6.19 *In $O(\xi q^2 + \mu q^3 e \log q)$ time one can test whether or not the group J in **6.2.2** satisfies $J \cong \mathrm{SU}(4, q)$ and, if so, then find an isomorphism $J \to \mathrm{SU}(4, q)$ together with the elements and subgroups in **6.2.2**(i-v).*

This is proved as in Corollary 3.20.

6.6.3   Total time and reliability

Part (i) of Theorems 1.1 and 1.1' is in Section 7; (ii) is **6.5.2**; (iii) and (iv) are in **6.5.3**; and (vi) is in **6.3.3**. For (vii), by tallying and recalling the recursive call we find that the *total time is*

$$O(d\{\xi[q^2 \log d + qd \log d] + \mu[q^3 ed \log q + q^2 ed^3 \log d]\}),$$

dominated by **6.2.1**, **6.2.2** and **6.5.2**. By **6.5.3**, the straight-line algorithms for (iii) and (iv) take $O(\mu[q^3 \log q + q^2 d^2] + d^5 \log q)$ and $O(\mu d^2 \log q)$ time, respectively.

Using the terminology in Theorems 1.1 and 1.1', we must replace q by $q^{1/2}$ and hence find that *the total time is $O(d\{\xi[q \log d + q^{1/2} d \log d] + \mu[q^{3/2} ed \log q + qed^3 \log d]\})$ for the entire algorithm in this section*, while the algorithm for Theorems 1.1(iii) and 1.1'(iii) takes $O(\mu[q^{3/2} \log q + qd^2] + d^5 \log q)$ time.

Reliability: $> 1/2$.

6.6.4   From $\mathrm{PSU}(4,q)$ **to** $\mathrm{P}\Omega^-(6,q)$

Finally, we discuss one of the ways to pass from the projective $\mathrm{PSU}(4,q)$–module to the projective $\mathrm{P}\Omega^-(6,q)$–module (also see **4.6.3**).

Write matrices generating the centralizer $Q^{\$}$ of a totally singular line Λ of $\mathrm{GF}(q^2)^4$, as well as matrices generating a subgroup $L \cong \mathrm{SL}(2,q^2)$ of the set stabilizer $\mathrm{SU}(4,q)_\Lambda$, and an element $h \in \mathrm{SU}(4,q)_\Lambda$ of order $q-1$ centralizing L.

This group $Q^{\$}$ is the group "Q" in **4.1.3**, and L acts on $Q^{\$}$ as $\Omega^-(4,q)$. The action of h makes $Q^{\$}$ a $\mathrm{GF}(q)$–space. Construct the desired vector space $\mathrm{GF}(q)^6$, together with its quadratic form, exactly as in **4.3.2–4.5.2**.

7 Proofs of Theorems 1.1 and 1.1′, and of Corollaries 1.2–1.4

In Sections 3–6, we worked with the assumption that the type, the size q of the underlying field, and the dimension d of G are known. With this assumption added to the hypotheses of Theorem 1.1, we constructed the *correct* output of Theorem 1.1 with probability greater than $1/2$; with probability less than $1/2$ the algorithm reported failure.

When the algorithm is run without the *a priori* knowledge of the type, q and d, we start the algorithm by determining at most 7 possibilities for these. Then we try all of these possibilities. We *assume* a certain type and values for d and q, and run the algorithm of the appropriate section. If an output is returned, we have to decide whether our current assumption for the type, d and q is correct. To this end, in this section we will construct a presentation for the classical group of that particular type and dimension, and check whether G satisfies this presentation. In particular, in addition *we verify the hypothesis of* Theorem 1.1 *that G is a classical simple group.*

7.1 Presentations

We describe algorithmically computable presentations for the classical simple groups.

7.1.1 The Curtis-Steinberg-Tits presentation

The Curtis-Steinberg-Tits presentation for G [Ca, pp. 176-188] is a presentation of length $O(d^4q^2)$. It arises from specific matrices given explicitly in [Ca, pp. 186–187, 271] for the cases $\mathrm{SL}(d,q)$, $\mathrm{Sp}(d,q)$, $\Omega(2m+1,q)$, $\Omega^\pm(2m,q)$, and implicitly by a specific procedure provided in [Ca, pp. 270–271] and implemented in [Gr]. As usual let $\mathbb{F}$ denote $\mathrm{GF}(q)$.

Example 1. Let $G = \mathrm{PSL}(d,q)$ with $d \geq 3$. Here and later, we let E_{ij} denote the $d \times d$ matrix all of whose entries are 0 except that the i,j-entry is 1. The generators are $x_{ij}(t) = I + tE_{ij}$, where $1 \leq i, j \leq d$ and $t \in \mathbb{F}$, with the following relations for all distinct 2–element sets $\{i,j\}, \{k,l\}$ from $\{1,\ldots,d\}$ and all $t, u \in \mathbb{F}$:

$$[x_{ij}(t), x_{kl}(u)] = \begin{cases} x_{il}(tu) & \text{if } i < j = k < l \\ x_{kj}(-tu) & \text{if } k < i = l < j \\ x_{jk}(-tu) & \text{if } j < i = l < k \\ x_{li}(tu) & \text{if } l < j = k < i \\ 1 & \text{otherwise} \end{cases}$$

140

These relations provide a presentation for the universal central extension of
PSL(d, q) [Ca, p. 190]. Additional relations are needed in order to obtain a
presentation with trivial center.

Example 2. Let $G = \mathrm{PSL}(2, q)$. Let U be the lower triangular Sylow p–
subgroup of G, let $\langle h \rangle$ be the group of diagonal matrices, and let $r \notin \langle h \rangle$ be an
involution normalizing $\langle h \rangle$. Then $N = \langle h, r \rangle$ is dihedral. A presentation for G
is obtained by starting with presentations for U and N, then giving the action
of h on U, and finally giving *all* relations of the form $w = uvu'$ with $w \in \langle h \rangle r$,
$u, u' \in U$, and $v \in r^{-1}Ur$ (there are $q - 1$ such relations, exactly one for each
nontrivial $u \in U$).

Example 3. When $G = \mathrm{Sp}(2m, q)$ with $2m \geq 4$, the standard basis of V is
labeled $e_{-1}, \ldots, e_{-m}, e_1, \ldots, e_m$ with all inner products 0 except $(e_{-i}, e_i) = 1 = -(e_i, e_{-i})$. Define E_{ij} as in Example 1 whenever $|i|, |j| \leq m$. The generators
are as follows, for $1 \leq i < j \leq m$ and $t \in \mathbb{F}$:

$$
\begin{aligned}
x_{ij}(t) &= I + t(E_{ij} - E_{-j,-i}), & x_{-i,-j}(t) &= I + t(E_{-i,-j} - E_{ji}), \\
x_{i,-j}(t) &= I + t(E_{i,-j} + E_{j,-i}), & x_{-ij}(t) &= I + t(E_{-ij} + E_{-ji}). \\
x_i(t) &= I + tE_{i,-i}, \quad \text{and} & x_{-i}(t) &= I + tE_{-ii}.
\end{aligned}
$$

All commutators of pairs of these generators can be easily computed provided
that the absolute values of the subscripts of a pair do not coincide. For example, $[x_{ij}(t), x_k(u)] = 1$ unless $j = k$, in which case it is $x_{i-k}(tu)x_i(-t^2u)$.
The relations of this sort provide a presentation for the universal central extension of PSp$(2m, q)$. Additional relations are needed in order to obtain a
presentation with trivial center.

Except for the unitary case, all classical groups of dimension ≥ 4 fall into
this pattern with matrices almost the same as the ones just written. A presentation for the unitary groups is given in [Gr], and is obtained using matrices
only slightly different from the ones above.

7.1.2   A shorter presentation

The presentation in the preceding section is shortened to one of length
$O(\log^2 |G|)$ in [BGKLP]. We will work with the field $\mathbb{F}$ in **2.3**: in [BGKLP]
this field is presumed available in terms of an irreducible polynomial of degree
e over GF(p). Then the following is contained in that paper:

Theorem 7.1 [BGKLP] *Each finite simple classical group G has a presenta-
tion of length $O(\log^2 |G|)$ (in fact, of length $O(\log |G|)$ if G is not unitary),*

with the possible exception of the unitary groups PSU$(3, q)$. *Moreover, such a presentation uses* $O(\log |G|)$ *generators, and can be obtained in* $O(\log^2 |G|)$ *time.*

The same statements hold for the group G in Theorem 1.1'.

Some explanations are relevant for our applications. Consider Example 1 in **7.1.1**. Let ρ be any generator of $\mathbb{F}^*$, and introduce generators called $y_{ij}(\rho^m)$ for i, j as before and $0 \le m < e$. The relations are the same as before using $t = \rho^m, u = \rho^n$, together with $[y_{ij}(\rho^m), y_{ij}(\rho^n)] = 1$ for all i, j, m, n and $[y_{ij}(\rho^m), y_{jk}(\rho^n)] = \prod_{r=0}^{e-1} y_{ik}(\rho^r)^{a_{mnr}}$ for $i < j < k$, where $\rho^m \rho^n = \sum_{r=0}^{e-1} a_{mnr} \rho^r$ with $a_{mnr} \in \mathrm{GF}(p)$. This is further shortened by expressing a_{mnr} in binary. Moreover, it is only necessary to consider those i and j that differ by 1. For this, and further shortenings of this presentation, see [BGKLP]. As before it is necessary to add relations to make the resulting group have center 1.

Example 3 is similar. Example 2 is somewhat different. More than commutators need to be used. Short presentations of $U\langle h \rangle$ and $\langle h, r \rangle$ are easy to write down. In particular, U has a presentation with pairwise commuting generators $y(\rho^m)$ as above. The relations $w = vuv'$ can be expressed as $ry(-1/d)ry(d)ry(-1/d) = h(d) \in \langle h \rangle$, where d is ρ^i, $0 \le i < e$ and $y(-1/d)$ is computed as in the preceding paragraph by writing $-1/d$ as a linear combination of the ρ^r, $0 \le r \le e-1$. This is cut to a presentation of length $O(\log q)$ by incorporating conjugation by h as a relation, $y(d)^h = y(\rho^2 d)$, thereby producing $\Theta(q)$ relations $w = vuv'$ from each such relation (here, $y(\rho^2 d)$ is computed as in the preceding paragraph).

Example 4. Let $G = \mathrm{PSU}(3, q)$. It is not known whether this has a presentation of length $O(\log q)$, although we expect that this is the case. The problem is similar to that in Example 2: this is a rank 1 group of Lie type, and hence has no nice presentation by commutator relations. Nevertheless, G has a presentation of length $O(q^3)$. Namely, let $\mathbb{F}^3$ have basis e_1, e_2, e_3 with $(e_1, e_3) = (e_2, e_2) = 1$, $(e_1, e_1) = (e_3, e_3) = (e_1, e_2) = (e_3, e_2) = 0$, and consider the matrices (6.16). Let U denote the group of q^3 matrices $x(c, d)$, and let $h = h(\rho)$ be a generator of the cyclic group of all $h(k)$ modulo scalar matrices. Short presentations of $U\langle h \rangle$ and $\langle h, r \rangle$ again are easy to write down. The only other required relations $w = vuv'$ are the $q^3 - 1$ relations (6.17) for all c, d satisfying $d + \bar{d} + c\bar{c} = 0$.

This can be cut to *a presentation of length* $O(q)$ by incorporating conjugation by h as a relation, $x(c, d)^h = x(c\rho^2/\bar{\rho}, \rho\bar{\rho}d)$, thereby producing $\Theta(q^2)$ relations $w = vuv'$ from each such relation. Note that Example 2 is obtained by restricting the above presentation to the generators $x(0, d)$ and r.

One further property of these presentations is

Proposition 7.2 *There is an $O(d^3 + d^2 \log q)$ time algorithm which, for each presentation in Theorem 7.1 and each $g \in G$, finds a straight-line program of length $O(d^2 \log q)$ from the generators to g.*

Proof. When $G = \mathrm{PSL}(d, q)$, $d \geq 3$, use left multiplication by the elementary matrices $x_{ij}(\rho^n)$ in order to row reduce g to a scalar matrix (note that row reduction expresses group elements in terms of the $x_{ij}(t)$, $t \in \mathbb{F}$ but we can then express these $x_{ij}(t)$ in terms of the $x_{ij}(\rho^n)$, $0 \leq n < e$). The same can be done when $G = \mathrm{PSL}(2, q)$ using $x(\rho^n)$ and $x(\rho^n)^r$. If we exclude the unitary groups, then pairs of opposite root groups (such as $x_{ij}(\mathbb{F})$ and $x_{-i,-j}(\mathbb{F})$, or $x_i(\mathbb{F})$ and $x_{-i}(\mathbb{F})$, in Example 3) can be used to row reduce g, eventually placing g into the Sylow p-subgroup that is the product of all positive root groups (all the $x_{ij}(\mathbb{F})$ and $x_i(\mathbb{F})$ in Example 3 with $i, j > 0$). In effect, this just row reduces within a subgroup $\mathrm{SL}(m, q)$; it is also a consequence of the Bruhat decomposition [Ca, 8.4]. The unitary case is similar, using generators behaving as described in [BGKLP] (based on [Gr]). □

Note the similarity to the proof of Theorem 1.1(iii,iv). We could have been a bit more efficient in the above proof and used left and right multiplication by generating matrices, thereby both row- and column-reducing g.

The same result holds for the groups G in Theorem 1.1′.

7.2   Completion of the proof of Theorem 1.1

In order to use the algorithms of Sections 3–6 and our presentations, first we need to know the (alleged) isomorphism type of G. We assume that $|G| > 4^{36}$ when $p \leq 3$.

7.2.1   Finding d, q and the type of G

Since we have assumed that $|G| > 4^{36}$ when $p \leq 3$, the $\mathrm{ppd}^{\#}(p; v)$-numbers we are about to use are defined and satisfy $v > 6$ when $q \leq 4$.

Choose $\lceil 32\sqrt{N} \rceil$ elements $g \in G$. Let v_1, v_2, where $v_1 > v_2$, be the two largest integers v such that some $\mathrm{ppd}^{\#}(p; v)$-number divides the order of at least one of the chosen g.

Let $e' := (v_1, v_2)$ and $u' := v_1/e'$. Based on the values of u' and e' we will use various tests for the type, dimension, and underlying field size of G.

First run the test for $\mathrm{PSL}(u', p^{e'})$ in Section 3 (including checking a presentation of $\mathrm{PSL}(u', p^{e'})$ using **7.2.2**). If it succeeds then $q := p^{e'}$ and $d := u'$.

If the previous test did not succeed then e' must be even.

If $u' \geq 3$ is odd run the tests for $\mathrm{PSU}(u', p^{e'/2})$ and $\mathrm{PSU}(u' + 1, p^{e'/2})$. If either succeeds let $q := p^{e'}$ and $d := u'$ or $u' + 1$, respectively.

Let $e := e'/2$, $q := p^e$ and $u := 2u'$.

Run the tests for $\mathrm{PSp}(u, q)$ if $u \geq 4$, $\mathrm{P\Omega}^+(u+2, q)$ if $u \geq 6$, $\mathrm{P\Omega}(u+1, q)$ if $u \geq 6$ and q is odd, and $\mathrm{P\Omega}^-(u, q)$ if $u \geq 8$. If any succeeds let $d := u$, $u + 2$, $u + 1$ or u, respectively.

Correctness: The possibilities for v_1 and v_2 in the various types of classical groups are summarized in the following table. As usual, we have written $q = p^e$; the values in the table can be read from the orders of the groups.

G	v_1	v_2
$\mathrm{PSL}(d, q)$, $d \geq 2$	ed	$e(d-1)$
$\mathrm{PSp}(2m, q)$, $m \geq 2$	$2em$	$e(2m-2)$
$\mathrm{P\Omega}^+(2m, q)$, $m \geq 4$	$e(2m-2)$	$e(2m-4)$
$\mathrm{P\Omega}(2m+1, q)$, $m \geq 3$	$2em$	$e(2m-2)$
$\mathrm{P\Omega}^-(2n, q)$, $n \geq 4$	$2en$	$e(2n-2)$
$\mathrm{PSU}(2m+1, q^{1/2})$, $m \geq 1$	$e(2m+1)$	$e(2m-1)$
$\mathrm{PSU}(2m, q^{1/2})$, $m \geq 3$	$e(2m-1)$	$e(2m-3)$
$\mathrm{PSU}(4, q^{1/2})$	$3e$	$2e$

In each case an element g whose order is divisible by a $\mathrm{ppd}^{\#}(p; v_i)$-number, $i = 1$ or 2, decomposes the underlying vector space as $V = V_{v_i} \oplus V_0$, where g acts on V_i as in **2.4**, **4.1.5**, **5.1.5**, or **6.1.5**; and $\dim V_0 \leq 4$. By considering the possible actions of g on V_0 we see that $|g|$ is a factor of $(p^{v_i} - 1)p(q+1)(q^3 - 1)$ (compare **4.2.1**, **5.2.1**, **5.2.2**, **3.2.1** and **6.2.1**).

The probability that an element of G has order of the stated form is at least $1/8d \geq 1/8\sqrt{N}$ by Lemma 2.5, **4.1.5**, **5.1.5** and **6.1.5**.

Reliability: $> 1 - 1/16$.

Time: $O(\sqrt{N}(\xi + \mu N^2))$ to find the integers v_i, since testing whether one g has order of the desired sort requires the recursive computation of $g^{\prod_{i=1}^{m}(p^i-1)}$ and $g^{p\prod_{i=1}^{m}(p^i-1)}$, for $1 \leq m \leq N/\log p$. Then a total of 7 tests are made using Sections 3–6 and the presentations.

Remarks. Once again we note that we do not see how to avoid starting with at least the value of p as part of the input, since there are exponentially many choices (as a function of N) for p, if the only information we have is $|G| \leq 2^N$.

With some extra work (but still in time polynomial in the input length), we could have narrowed the possibilities for the type of G down to 3. This can be done by searching for group elements whose order is divisible by two appropriately chosen ppd$^{\#}(p; x)$-numbers.

7.2.2   Verifying a presentation

We have constructed an *alleged* homomorphism $\lambda \colon G \to \mathrm{PSL}(V)$ with concrete image group $C \leq \mathrm{PSL}(V)$. We have the required form or forms on V. Thus, we can quickly perform a change of basis so that the forms behave the same as the ones in [Ca, pp. 186–187, 271] do on the standard basis of $V = \mathbb{F}^d$.

Write a short presentation $\langle X \mid R \rangle$ of C in $O(d^4 \log^2 q)$ time ($O(q^{1/2})$ time for $\mathrm{PSU}(3, q^{1/2})$). This means that we construct a map $\varphi \colon X \to C$ such that, if $F(X)$ is the free group with basis X and if $N = \langle R^{F(X)} \rangle$, then φ induces an isomorphism $\hat{\varphi} \colon F(X)/N \to C$.

Use Theorem 1.1(iv) to find $\mathcal{S}^{**} := X\varphi\lambda^{-1} \subseteq G$. For each word $w(x_1, \ldots) \in R$ (where $x_1, \ldots \in X$), find $w(x_1\varphi\lambda^{-1}, \ldots)$ and test whether this is 1 in G. If so, then $C \cong \langle \mathcal{S}^{**} \rangle$; otherwise $C \not\cong G$.

Finally, we need to test that $G = \langle \mathcal{S}^{**} \rangle$ by verifying that $\mathcal{S} \subseteq \langle \mathcal{S}^{**} \rangle$. Use Proposition 7.2 to write straight-line programs from $X\varphi$ to $\mathcal{S}\lambda$, and apply these programs starting with $X\varphi\lambda^{-1} = \mathcal{S}^{**}$ in order to obtain $\mathcal{S}$. Of course, if $\mathcal{S} \subseteq \langle \mathcal{S}^{**} \rangle$ is false then $C \not\cong G$.

Once we know that $C \cong G$, it follows from Sections 3–6 that λ is, indeed, an isomorphism.

Time: Temporarily ignore the case $\mathrm{PSU}(3, q^{1/2})$. It takes $O(d^4 \log^2 q)$ time to write the presentation, $O(d^2 \log q \cdot [\mu d^2 \log q])$ to apply λ^{-1} to obtain $\mathcal{S}^{**}$ (by Propositions 3.17, 4.22, 5.17 and 6.14), and $O(\mu d^4 \log^2 q)$ to check that the relations hold in $\mathcal{S}^{**}$. Also, $\mathcal{S}\lambda$ can be computed in $O(|\mathcal{S}|\mathcal{T})$ time, where $\mathcal{T}$ is the time requirement in the appropriate one of Propositions 3.18, 4.23, 5.18 and 6.15. The straight-line programs from $X\varphi$ to the members of $\mathcal{S}\lambda$ are obtained in $O(|\mathcal{S}|(d^3 + d^2 \log q))$ time by Proposition 7.2. It takes $O(|\mathcal{S}|\mu d^2 \log q)$ time to multiply out the mirrored straight-line programs of length $O(d^2 \log q)$ in G. Adding this time requirement to the time used in Sections 3–6 and in **2.3**, we obtain the following total time to handle G, *assuming* that G is isomorphic to the stated group. Note that we have used the assumption $\xi \geq \mu|\mathcal{S}|$ (cf. **2.2.2**) and the observation that $d^2 \log q$ is $O(\mu)$ (since $\mu \geq N \geq \log |G|$ and $\log |G|$ is $\Theta(d^2 \log q)$) to simplify some of the expressions.

$$
\begin{array}{ll}
\text{PSL}(d,q) & O(\xi q(d + \log q)d \log d + \mu q d^4 \log^2 q \log d \\
& \quad + |\mathcal{S}|d^5 \log q) \\[1ex]
\text{P}\Omega^\varepsilon(d,q),\ d \geq 7 & O(\xi q(d + \log q)d \log d + \mu q d^4 \log^2 q \log^3 d \\
& \quad + |\mathcal{S}|(d^5 \log q + \mu q d^2 \log q)) \\[1ex]
\text{PSp}(d,q),\ d \geq 4 & O(\xi[q + d]d \log d \log q + \mu d^4 \log^2 q \log d \quad\quad (7.3)\\
& \quad + |\mathcal{S}|d^5 \log q) \\[1ex]
\text{PSU}(d,q^{1/2}),\ d \geq 4 & O(\xi[q d \log d + q^{1/2} d^2 \log d] \\
& \quad + \mu[q^{3/2}d^2 \log^2 q + q d^4 \log q \log d] \\
& \quad + |\mathcal{S}|(d^5 \log q + \mu q^{3/2} d^2 \log q)
\end{array}
$$

Again using that $d^2 \log q$ is $O(N)$, the time requirement in (7.3) is $O(\xi q N \log N + \mu[q N^2 \log^3 N + q^{3/2} N^2] + |\mathcal{S}|(N^{5/2} + \mu q^{3/2} N))$, for G of *known* isomorphism type. However, we need to determine at most seven possibilities of isomorphism types for G in **7.2.1** and test all of these. This adds a term $\mu N^{5/2}$ to the overall timing of the algorithm. The reliability of the algorithm is $> 1/2$, since success depends only on how the algorithm fares in *one* of at most seven runs, when the correct isomorphism type is assumed.

Finally, we test (if necessary) whether $G \cong \text{PSU}(3, q^{1/2})$. For this, recall from **7.1.2** that it takes $O(q^{1/2})$ time to write the required presentation of length $O(q^{1/2})$, after which the timing is as before: $O(\xi q + \mu q^{3/2} \log q)$, including the time to verify the presentation.

Thus, *the total time is* $O(\xi q N \log N + \mu[q^{3/2} N^2 \log^3 N + N^{5/2}] + |\mathcal{S}|[\mu q^{3/2} N + N^{5/2}])$, as required in the statement of the theorem. This completes the proof of Theorem 1.1. The proof of Theorem 1.1$'$ is completed in exactly the same way. $\square$

7.2.3   Monte Carlo simplicity test

It is difficult to imagine any efficient Las Vegas test for whether a given black box group is *not* isomorphic to a *given* simple group. However, our algorithm does provide an *efficient Monte Carlo test for whether a given black box group G is or is not isomorphic to a classical group of known characteristic.* Namely, just run the algorithm for Theorem 1.1. If the algorithm produces an output then an isomorphism is obtained. If the algorithm reports failure then, with probability $> 1/2$, G is not isomorphic to a classical simple group.

7.3   Corollaries 1.2–1.4

Corollary 1.2 is a consequence of the remarks following the statement of Theorem 1.1'.

Proof of Corollary 1.3. We are given $b \in B$. Test whether or not $b \in G$ by running the algorithm in Proposition 3.18, 4.23, 5.18 or 6.15 using b in place of g. This either produces a straight-line program from $\mathcal{S}^*$ to b, or reports failure, in which case $b \notin G$. Similarly, $b \in \mathrm{N}_B(G)$ can be tested by checking whether $x^b \in G$ for all $x \in \mathcal{S}^*$.

Finally, we have to test whether $b \in G \cdot \mathrm{C}_B(G)$. The only subcase we have left to handle is when b normalizes G, but $b \notin G$. Run the aforementioned algorithm for b, to the point when it replaces b (by multiplications with elements of G) by an element of $\mathrm{C}_{\mathrm{N}_B(G)}(Q) = \mathrm{Z}(Q)\mathrm{C}_B(G)$ (or reports failure at an earlier stage). Here, the latter equation follows from the fact that no outer automorphism of G can centralize Q. $\square$

Remarks. The same result holds if G is a classical linear group, as in Theorem 1.1'.

Of course, the above argument proves slightly more. Namely, for *any* black box group G for which there is a generating set $\mathcal{S}^*$ and an oracle which finds any given element of G by a straight-line program from $\mathcal{S}^*$, whenever G is a subgroup of a black box group B there are membership tests deciding whether any given element of B lies in G or $\mathrm{N}_B(G)$.

Proof of Corollary 1.4. Compute $O(N)$ generators for $G''' = G_0$ by a Monte Carlo algorithm, obtaining such generators with probability $> 1 - 1/100$. This can be done in $O(\mu[N^2 \log N + |\mathcal{S}| \log N])$ time [BCFLS]; in fact, $O(\mu[N \log^4 N + |\mathcal{S}| \log N])$ suffices [Ser2]. Then run the algorithm of Theorem 1.1; in particular, this verifies that the computation of G''' was correct.

Now we have $G_0 = \langle \mathcal{S}^* \rangle$, λ, and the vector space V underlying G_0. In most cases, G acts projectively on V as a group of semilinear maps. When this occurs, we imitate the algorithms that found elements of G_0 (in **3.5.2**, **4.5.2**, **5.5.2**, **6.5.2**). Namely, let $g \in G$. We have certain points $Q(\beta_i)$ (or $T(\beta_i)$), their labels $\langle v_i \rangle$, and the labels $\langle w_i \rangle$ of their images under g. Hence, the desired transformation $g\lambda$ must send $\langle v_i \rangle$, to $\langle w_i \rangle$.

If $g\lambda$ is a linear map then we can proceed exactly as before. However we need to discuss the possibility that $g\lambda$ is only semilinear. For this purpose we will consider just the case of the orthogonal groups; the other types are similar but slightly simpler.

In **4.2.2**(v) we introduced elements $j(\beta)$, $j(\beta') \in J$ such that the corresponding points $Q(\beta) = Q(\alpha)^{j(\beta)}$, $Q(\beta') = Q(\alpha)^{j(\beta')}$ have respective labels $\langle v_{d-1} \rangle$, $\langle v_d + m v_{d-1} \rangle$. We may assume that $m = 1$ and that the label $\langle w_{d-1} \rangle$ of $Q(\beta_{d-1})^g$ has been rescaled so that $Q(\beta')^g$ has label $\langle w_{d-1} + w_d \rangle$. Now we introduce yet another point if q is not a prime: $Q(\beta'') := Q(\alpha)^{j(\beta'')}$, $j(\beta'') \in J$, with label $\langle v_d + \rho v_{d-1} \rangle$ (cf. **2.3**). Then $Q(\beta'')^g$ has label $\langle w_d + \rho'' w_{d-1} \rangle$ for some $\rho'' \in \mathbb{F}$, and $\rho \mapsto \rho''$ defines the required field automorphism.

Note that this procedure can be used whenever $g \in G$ acts projectively on V.

The only cases in which G might not act on V are as follows:

(1) G_0 is $\mathrm{PSL}(d,q)$, and G contains an element interchanging V with its dual space; and

(2) G_0 is $\mathrm{P\Omega}^+(8,q)$, and G contains an element sending V to one of the half-spin modules for $\Omega^+(8,q)$.

In these cases G acts projectively on the direct sum of 2 or 3 vector spaces, where G permutes these spaces among themselves (projectively). A matrix representing each element of $\mathcal{S} \cup \mathcal{S}^*$ in this action can be obtained as before:

(1) G_0 is $\mathrm{PSL}(d,q)$. Starting with G_0 we obtain a vector space V. There is a second possible vector space, namely the dual V^*, and it is straightforward to find it. Now G acts on $V \oplus V^*$, and we can find matrices as before.

(2) G_0 is $\mathrm{P\Omega}^+(8,q)$. Starting with G_0 we obtain a vector space V. This time there are two other choices for V, equivalent to V under $\mathrm{Aut}G_0$ but producing nonisomorphic projective G_0–modules. Instead of going through spinors to construct these modules, we proceed as follows.

Each $g \in \mathcal{S}$ permutes the G–conjugacy class of the group Q appearing in **4.1.3**. Hence, g will send V to one of the relevant three vector spaces. Now G acts on the direct sum of at most three 8–spaces, and we can find matrices as before.

This completes the proof of Corollary 1.4. $\square$

Remark. Again the same result holds if G_0 is a classical linear group, as in Theorem 1.1'.

8 Permutation group algorithms

8.1 Bases, strong generating sets, and Schreier trees

Fundamental data structures for computing with permutation groups were introduced by Sims in [Si1, Si2]. A *base* for a permutation group $G \leq \mathrm{Sym}(\Omega)$ of degree n is a sequence $B = (\beta_1, \beta_2, \ldots, \beta_M)$ of points from Ω such that the pointwise stabilizer $G_B = 1$. The *point-stabilizer chain* of G relative to B is the chain of subgroups

$$G = G^{(1)} \geq G^{(2)} \geq \cdots \geq G^{(M+1)} = 1,$$

where $G^{(i)} = G_{(\beta_1, \ldots, \beta_{i-1})}$. The base B is called *nonredundant* if there is strict inclusion $G^{(i)} > G^{(i+1)}$ for all $1 \leq i \leq M$. If B is nonredundant then $2^{|B|} \leq |G| \leq n^{|B|}$ or, equivalently, $(\log |G|)/(\log n) \leq |B| \leq \log |G|$. A *strong generating set* (SGS) for G relative to B is a set $\mathcal{S}$ of generators of G with the property that

$$\langle \mathcal{S} \cap G^{(i)} \rangle = G^{(i)} \text{ for } 1 \leq i \leq M + 1.$$

Let $B = (\beta_1, \ldots, \beta_M)$ be a base of the group G, let $G = G^{(1)} \geq G^{(2)} \geq \cdots \geq G^{(M+1)} = 1$ be the corresponding point-stabilizer chain, and let R_i denote a transversal for $G^{(i+1)}$ in $G^{(i)}$, $1 \leq i \leq M$. Such a transversal can be computed from the SGS by a standard orbit computation of $\beta_i^{G^{(i)}}$, keeping track of group elements sending β_i to the points of the orbit. Each $g \in G$ can be written uniquely in the form

$$g = r_M r_{M-1} \cdots r_2 r_1, \ r_i \in R_i. \tag{8.1}$$

The process of factoring g in this form is called *sifting* or *stripping*. Note that $|G|$ is easily computable as $|G| = \prod_1^M |R_i|$.

In practical computation, the transversals R_i usually are not computed and stored explicitly; rather, they are encoded in a Schreier-tree data structure. Suppose that a base B and an SGS $\mathcal{S}$ for G relative to B are given. A *Schreier-tree data structure* for G is a sequence of pairs (S_i, T_i) called *Schreier trees*, one for each base point β_i, $1 \leq i \leq M$, where T_i is a directed labeled tree with all edges directed toward the root β_i, and with edge-labels selected from the set $S_i := \mathcal{S} \cap G^{(i)} \subseteq G^{(i)}$. The nodes of T_i are the points of the orbit $\beta_i^{G^{(i)}}$. The labels satisfy the condition that $\gamma^h = \delta$ for each directed edge from γ to δ with label h. If γ is a node of T_i, then the sequence of the edge-labels along the path from γ to β_i in T_i is a word in the elements of S_i such that the product of these permutations moves γ to β_i. Thus each Schreier tree (S_i, T_i) defines *inverses* of the elements of a transversal for $G^{(i+1)}$ in $G^{(i)}$.

The reason for storing the inverses of transversal elements is that the sifting procedure uses these inverses. For fast sifting, it is crucial that the depths of trees in the Schreier-tree data structure are small. In fact, the SGS construction algorithm in [BCFS] constructs Schreier trees such that the sum of their depths is $O(\log|G|)$. We call a Schreier-tree data structure *shallow* if the depth of each tree is at most $2\log|G|$. A shallow Schreier-tree data structure supports membership testing in $O(nM\log|G|)$ time.

Lemma 2.2 in [BCFS] states that, given $G = \langle S \rangle \le \mathrm{Sym}(\Omega)$ and $\alpha \in \Omega$, group elements $g_1, \ldots, g_k$ (for some $k \le \log|G|$) can be computed in deterministic $O(n|S|\log^2|G|)$ time such that the orbits α^G and $\alpha^{\langle g_1, \ldots, g_k \rangle}$ are equal, and the breadth-first-search tree for α^G, using the group elements $g_1, \ldots, g_k, g_1^{-1}, \ldots, g_k^{-1}$, has depth at most $2k \le 2\log|G|$. This implies that, given a nonredundant base $B = (\beta_1, \ldots, \beta_M)$ and an SGS S relative to B for a group G, we can replace S in $O(nM|S|\log^2|G|)$ deterministic time by another SGS T relative to B such that $|T| \le 2M\log|G|$ and the Schreier-tree data structure obtained from T is shallow. Therefore, we will assume that all Schreier-tree data structures computed in our algorithms are shallow.

8.2   Nearly linear time algorithms

Measuring the efficiency of a permutation group algorithm solely as a function of the input length often does not indicate properly the practical performance of the algorithm. Given $G = \langle S \rangle \le S_n$, the running time often does not only depend on the input length $|S|n$, but also on the order of G. Moreover, in groups of current interest for implementations, it frequently happens that the degree of G is in the tens of thousands or even higher, so even a $\Theta(n^2)$ algorithm may not be practical. On the other hand, $\log|G|$ is often small, bounded from above by a polylogarithmic function of n. Therefore, a recent trend is to search for algorithms with running time of the form $O(n|S|\log^k|G|)$. These notions are formalized in the following way.

With reference to some constant c, an (infinite) family $\mathcal{G}$ of permutation groups is called a family of *small-base groups* if all $G \in \mathcal{G}$ of degree n admit bases of size $O(\log^c n)$. Equivalently, there is a constant c' such that $\log|G| = O(\log^{c'} n)$ for each $G \in \mathcal{G}$ of degree n.

In particular, the classical simple groups, considering all of their permutation representations, comprise a small-base family (with $c = 2$):

Lemma 8.2 [Co] *Let G be a simple classical group, defined on a vector space of dimension d over $\mathrm{GF}(q)$. Suppose that G acts faithfully on a set Ω of size n. Then, for some constant c', n is at least $c'q^{d-1}$ if G is $\mathrm{PSL}(d, q)$ or symplectic, at least $c'q^{d-2}$ if G is orthogonal, and at least $c'q^{d-3/2}$ if G is unitary.*

We call a permutation group algorithm a *nearly linear time algorithm* if its running time for any $G = \langle S \rangle \le S_n$ is $O(n|S| \log^k |G|)$. The name is justified by the fact that, if G is a member of a small-base family then the running time is a nearly linear, $O(n|S| \log^{c''}(n|S|))$, function of the input length. We will require the following algorithms of this sort:

Theorem 8.3 *There are Monte Carlo nearly linear time algorithms which, when given $G \le S_n$, find the following:*

 (i) *[BCFS] A base, strong generating set, and a shallow Schreier-tree data structure for G;*

 (ii) *Given a homomorphism $\varphi\colon G \to S_n$ specified by the images of generators, data structures which enable the nearly linear time computation of $\varphi(g)$ for any $g \in G$ and a preimage of any $g \in \varphi(G)$ (this is a consequence of (i));*

 (iii) *[BeS] A composition series $G = N_1 \rhd N_2 \rhd \cdots \rhd N_m = 1$ and, for each $1 \le i \le m - 1$, a homomorphism $\varphi_i\colon N_i \to S_n$ with $\ker \varphi_i = N_{i+1}$.*

Nearly linear algorithms are not merely theoretical advances: a large part of the permutation group library in **GAP** is based on implementations of nearly linear algorithms.

8.3 Permutation groups as black box groups

Suppose that a base $B = (\beta_1, \ldots, \beta_M)$, a strong generating set S with respect to B, and a Schreier-tree data structure $\mathcal{ST} = \{(S_i, T_i) \mid 1 \le i \le M\}$ are given for some $G \le \mathrm{Sym}(\Omega)$. Let t be the sum of the depths of these M Schreier trees. As indicated in **8.1**, we can assume that the Schreier trees are shallow, so t is $O(\log^2 |G|)$. We shall also assume that the SGS S is closed under taking inverses (this condition is automatically satisfied if the shallow Schreier trees were constructed by the method of Lemma 2.2 in [BCFS]).

Any $g \in G$ can be written uniquely in the form $g = r_M r_{M-1} \cdots r_1$, where the r_i are elements of the transversals whose inverses were coded by $\mathcal{ST}$. Each such inverse can be written as a word in the strong generators $\mathcal{S}$, following the path in the appropriate Schreier tree. Taking the inverse of this word and using the fact that $\mathcal{S} = \mathcal{S}^{-1}$, we obtain the r_i, and so g, as a word in $\mathcal{S}$ in a well-defined way. The length of the word representing g is at most t. We call this word the *standard word representing g*.

Lemma 8.4 *In deterministic $O(t|B|)$ time, given an injection $f: B \to \Omega$, it is possible to find a standard word representing some $g \in G$ with $B^g = f(B)$ or to determine that no such element of G exists.*

Proof. The element $g \in G$ with the required property can be obtained by a modification of the standard sifting procedure for permutations. Let $G = G^{(1)} \geq \cdots \geq G^{(M+1)} = 1$ be the point-stabilizer chain corresponding to B. If $f(\beta_1)$ is in the orbit $\beta_1^{G^{(1)}}$ then, taking the sequence of edge-labels along the path from $f(\beta_1)$ to β_1 in the first Schreier tree, we obtain a word w_1 such that $f(\beta_1)^{w_1} = \beta_1$. Hence, the function $f_2: B \to \Omega$ defined by $f_2(\beta_i) := f(\beta_i)^{w_1}$ fixes β_1. If $f_2(\beta_2) \in \beta_2^{G^{(2)}}$ then the sequence of edge-labels along the path from $f_2(\beta_2)$ to β_2 in the second Schreier tree defines a word w_2 such that the function $f_3: B \to \Omega$ defined by $f_3(\beta_i) := f(\beta_i)^{w_1 w_2}$ fixes the first two base points. Continuing this process, we obtain a word $w = w_1 w_2 \cdots w_M$ such that $f(\beta_i)^w = \beta_i$ for all $\beta_i \in B$. Taking the inverse of w, we get the standard word representing the desired $g \in G$. If, for some $1 \leq i \leq M$, $f_i(\beta_i) \notin \beta_i^{G^{(i)}}$, then we conclude that there is no $g \in G$ with $f(B) = B^g$. □

In the case when the SGS was computed by a Monte Carlo algorithm, it is possible that the preceding algorithm returns an incorrect answer, since it presupposes the correctness of our base, SGS and Schreier-tree data structure.

Now *we consider G as a black box group H*. The elements of H are defined to be the standard words representing the elements of G; these are strings over the alphabet $\mathcal{S}$, of length at most t. Of course, we can write the elements of H as 0-1 strings of uniform length N: $\mathcal{S}$ can be coded by $\lceil \log(|\mathcal{S}| + 1) \rceil$–long 0-1 sequences for the numbers $1, 2, \ldots, |\mathcal{S}|$, and every standard word can be padded by 0's to length t. Since we use shallow Schreier trees, $|\mathcal{S}|$ and t are $O(\log^2 |G|)$, and so N is $O(\log^2 |G| \log\log |G|)$. As customary in the analysis of permutation group algorithms, we assume that small numbers can be read in $O(1)$ time, and therefore we shall ignore the factors $\log\log |G|$ above. (This convention is consistent with the one that the product of two permutations in S_n can be computed in $O(n)$ time; it has also been used tacitly, e.g., in the proof of Lemma 8.4.)

Formally, we have an isomorphism $\psi\colon G \to H$ with the following properties. Each $g \in G$ defines an injection $f\colon B \to \Omega$ by $f(\beta_i) := \beta_i^g$, and Lemma 8.4 can be used to compute $g\psi$ in $O(\log^2 |G| \log |G|)$ time. Conversely, given $h \in H$, $h\psi^{-1}$ can be computed in $O(n \log^2 |G|)$ time, by multiplying out the product of the elements of h as a permutation.

Each $h \in H$ is represented by a unique string, so comparison of group elements can be performed in $O(\log^2 |G|)$ time. In order to take the product of $h_1, h_2 \in H$, we concatenate these two words, and define a function $f\colon B \to \Omega$ by $f(\beta_i) := \beta_i^{h_1 h_2}$. Then the standard word representing $h_1 h_2$ can be obtained by Lemma 8.4. This procedure runs in $O(\log^3 |G|)$ time. Similarly, to take the inverse of some $h \in H$, we take the inverse of the word h. This defines an injection $f\colon B \to \Omega$, and again we use Lemma 8.4 in $O(\log^3 |G|)$ time. Hence, *we have a black box oracle which performs the black box group operations in* $O(\log^3 |G|)$ *time.* Note, however, that this oracle can give incorrect answers if our base, SGS or Schreier-tree data structure was incorrect.

We can also construct uniformly distributed, independent random elements of B in $O(\log^2 |G|)$ time. Namely, we choose a uniformly distributed random point from each orbit $\beta_i^{G^{(i)}}$, and concatenate the corresponding transversal elements.

In particular, if $G \leq S_n$ is a member of a small-base family, then, in the notation of Theorem 1.1, ξ and μ are $O(\log^c n)$ for some constant c.

We shall apply Theorem 1.1 to $G\psi$ for permutation groups G which are isomorphic to classical matrix groups. The length of strings representing the elements of $G\psi$ and the cost of the black box group operations was discussed above. However, there are two minor technical points which imply that we cannot use the estimate $O(\log^2 |G|)$ for ξ, as stated above. First, in **2.2.2** we assumed $\xi \geq \mu|\mathcal{S}|$. Since $|\mathcal{S}|$ is $O(\log^2 |G|)$ and μ is $O(\log^3 |G|)$, this requires that ξ has to be estimated by $O(\log^5 |G|)$. Second, since during the recursive calls we need random elements in subgroups of $G\psi$ where Theorem 2.2 must be used, the estimate $O(\log^8 |G|)$ is required for ξ. The exponent of $\log|G|$ can be lowered in these instances as well as at other parts of the algorithms dealing with permutation groups; however, in this paper, our goal is just to establish the nearly linear running time, and we do not try to optimize the exponent of $\log|G|$.

8.4   Proof of Corollary 1.6

Given $G = \langle T \rangle \le S_n$, we use Theorem 8.3(i) to find a base, strong generating set, and shallow Schreier-tree data structure for G. The order of a simple group leaves at most two possibilities for the isomorphism type of G; in particular, if $|G| > 20160$ then $|G|$ uniquely determines q [Ar].

By **8.3** we can consider G as a black box group $H = G\psi$, and apply Theorem 1.1 for $G\psi$. The output of the algorithm in Theorem 1.1 contains a generating set $\mathcal{S}^*$ of $G\psi$ and a monomorphism $\lambda\colon G\psi \to \mathrm{PSL}(d, q)$, for the appropriate d, q, defined by giving the λ-images of the elements of $\mathcal{S}^*$. The composite mapping $\psi\lambda\colon G \to \mathrm{PSL}(d, q)$ is the desired monomorphism from G to $\mathrm{PSL}(d, q)$, and it is defined in terms of the images of the elements of $\mathcal{S}^*\psi^{-1}$.

As already noted, this algorithm is only Monte Carlo, since the original SGS computation may have returned an incorrect answer. *Now we upgrade this algorithm to Las Vegas.* Currently, we *know* that $|G| \ge |C|$ for some simple group $C \le \mathrm{PSL}(d, q)$, since the SGS construction made $|C|$ elements of G available as products of transversal elements of the form (8.1), and we want to *prove* that $G \cong C$. Recall (cf. **7.2**) that the algorithm of Theorem 1.1 also constructed a subset $\mathcal{S}^{**}$ of $G\psi$, and a presentation of C in terms of $\mathcal{S}^{**}$. In **7.2** it was also checked that $\mathcal{S}^{**}$ satisfies this presentation, but this check used the black box multiplication oracle, which we cannot rely on in this verification. So this time we check that:

(i) $\mathcal{S}^{**}\psi^{-1}$ satisfies the presentation for C; and

(ii) $T \subset \langle \mathcal{S}^{**}\psi^{-1} \rangle$.

Properties (i) and (ii), combined with the fact that $|G| \ge |C|$, prove that $G \cong C$, and so the SGS construction is correct.

Time: By Theorem 8.3(i), the initial SGS construction runs in nearly linear time. The time for the application of Theorem 1.1 can be read out of the appropriate line of (7.3). By **8.3**, μ, ξ, d and $\log q$ can be estimated from above by polynomial functions of $\log |G|$, and Lemma 8.2 implies that the polynomials of q occurring in the running time can be estimated from above by n. Hence the application of Theorem 1.1 runs in nearly linear time.

Finally, we estimate the time required for the upgrading of the algorithm to Las Vegas.

(i) Since the 3–dimensional unitary groups have been excluded, we have that $|\mathcal{S}^{**}|$ is $O(\log |G|)$ and the length of the presentation for C is $O(\log^2 |G|)$.

Hence the set $\mathcal{S}^{**}\psi^{-1}$ of permutations can be computed in $O(n\log^3|G|)$ time and the check that $\mathcal{S}^{**}\psi^{-1}$ satisfies the presentation requires $O(n\log^2|G|)$ time.

(ii) By Lemma 8.4, $\mathcal{T}\psi$ can be obtained in $O(|\mathcal{T}|\log^3|G|)$ time. Using the appropriate one of Propositions 3.18, 4.23, 5.18 and 6.15, and then Lemma 8.2 to replace the polynomial of q occurring in the running time by the degree n, we see that $\mathcal{T}\psi\lambda$ is computed in $O(|\mathcal{T}|n\log^5|G|)$ time. By Proposition 7.2, straight-line programs from $\mathcal{S}^{**}\lambda$ to $\mathcal{T}\psi\lambda$, of total length $O(|\mathcal{T}|\log|G|)$, can be computed in $O(|\mathcal{T}|\log^{3/2}|G|)$ time. Then the check that $\mathcal{T}\subset\langle\mathcal{S}^{**}\psi^{-1}\rangle$ can be done by mirroring these straight-line programs, starting from $\mathcal{S}^{**}\psi^{-1}$, in $O(n|\mathcal{T}|\log|G|)$ time. $\square$

8.5   Proofs of Theorem 1.7 and Corollary 1.8

Proof of Theorem 1.7.   Let $G = \langle\mathcal{T}\rangle \le S_n$ be a permutation group. Compute a base and strong generating set, and a composition series $G = N_1 \rhd N_2 \rhd \cdots \rhd N_m = 1$ by the nearly linear Monte Carlo algorithms in Theorem 8.3. The composition series algorithm also provides homomorphisms $\varphi_i \colon N_i \to S_n$ with $\ker\varphi_i = N_{i+1}$, for $1 \le i \le m-1$. We also compute strong generating sets for all N_i with respect to the base of G. We will verify the correctness of the base for G and the strong generating sets for the subgroups N_i by induction on $i = m, m-1, \ldots, 1$.

Suppose that we already have verified an SGS for N_{i+1}. We compute a base, SGS, and shallow Schreier-tree data structure in the image $N_i\varphi_i$, which is a subgroup of S_n and is allegedly isomorphic to a simple group. Our first goal is to obtain a presentation of length $O(\log^2|N_i\varphi_i|)$ for $N_i\varphi_i$, using a generating set $\mathcal{S}_i^{**}$ such that straight-line programs of length $O(\log|N_i\varphi_i|)$ from $\mathcal{S}_i^{**}$ to any given element of $N_i\varphi_i$ can be obtained in $O(n\log^5|N_i\varphi_i|)$ time.

The order $|N_i\varphi_i|$ gives at most two possibilities for the isomorphism type of $N_i\varphi_i$ [Ar]. Also, we have a Monte Carlo verification that G has no exceptional composition factors, since none of the orders of exceptional groups can coincide with the order of any other simple group. Hence we have at most two candidate simple groups for the isomorphism type of $N_i\varphi_i$, and in the ambiguous cases we try both possibilities. Also, if $|N_i\varphi_i| > 8!/2$ then $|N_i\varphi_i|$ determines whether $N_i\varphi_i$ is of Lie type, and if it is, determines its characteristic.

If $N_i\varphi_i$ is cyclic, it is trivial to get the desired presentation and straight-line programs. If it is alternating, say $N_i\varphi_i \cong A_k$, then [Mo2] describes a nearly linear time Monte Carlo algorithm to get the natural permutation representation of $N_i\varphi_i$, acting on k points. From this natural representation, a presentation of length $O(\log^2 |N_i\varphi_i|)$ is easily computable [Carm]. This presentation uses two generators: in the natural representation, one of them acts as a 3-cycle, while the other one as a $k-2$-cycle or the product of a $k-2$-cycle and a transposition, depending on the parity of k. It is easy to write straight-line programs of length $O(k \log k)$ to any element of $N_i\varphi_i$ from these generators [BLNPS]. Finally, if $N_i\varphi_i$ is a classical group, then we apply Corollary 1.6 to obtain the desired presentation and straight-line programs.

After this preprocessing, the correctness of the SGS for N_i and the correctness of φ_i can be proved as in the proof in **8.4**. Namely, let $\mathcal{T}_i$ be the set of generators of N_i computed by the composition series algorithm (Theorem 8.3(iii)). We check that (i) $N_{i+1} \trianglelefteq N_i$; (ii) $\langle \mathcal{S}_i^{**}\varphi_i^{-1}\rangle N_{i+1}/N_{i+1}$ satisfies the presentation computed for $N_i\varphi_i$; and (iii) $\mathcal{T}_i \subset \langle \mathcal{S}_i^{**}\varphi_i^{-1}\rangle N_{i+1}$, where $h\varphi_i^{-1}$ denotes a lift (i.e., an arbitrary preimage) of any given element $h \in N_i\varphi_i$.

(i) Conjugate the generators of N_{i+1} by the elements of $\mathcal{T}_i$ and check that the resulting permutations are in N_{i+1}. (Since the correctness of N_{i+1} is already known, membership testing is available for that group.)

(ii) Multiply out the relators that were written in terms of $\mathcal{S}_i^{**}$, using the permutations in $\mathcal{S}_i^{**}\varphi_i^{-1}$ as generators, and check that the resulting permutations are in N_{i+1}.

(iii) Write straight-line programs from $\mathcal{S}_i^{**}$ to $\mathcal{T}_i\varphi_i$, and for each $t \in \mathcal{T}_i$, execute the appropriate straight-line program, starting from $\mathcal{S}_i^{**}\varphi_i^{-1}$. This produces some $t^* \in \langle \mathcal{S}_i^{**}\varphi_i^{-1}\rangle$, and we check that $t^*t^{-1} \in N_{i+1}$.

This verifies (i-iii). At the end of the induction, we know a correct SGS for the group $N_1 = \langle \mathcal{T}_1\rangle$, output by the composition series algorithm. We know that $N_1 \leq G$, since the Monte Carlo algorithms in Theorem 8.3 never return generators for subgroups of G which are not elements of G. As a final check, we sift the elements of the original generating set $\mathcal{T}$ in N_1, verifying that $G \leq N_1$.

Essentially the same argument as in **8.4** shows that this algorithm runs in nearly linear time. We note that we have to require that calls to Corollary 1.6 fail with probability $< 1/(2\log|G|)$, since during the induction, $m \leq \log|G|$ such calls may be made; however, this multiplies the running time only by a $\log\log|G|$ factor. □

Proof of Corollary 1.8.   The following result is contained in [BGKLP, Sec. 8]. *If each composition factor H_i of the finite group G has a presentation of length $O(\log^C |H_i|)$ for some $C \geq 2$, then G has a presentation of length $O(\log^{C+1} |G|)$.* We shall indicate how the presentation described in [BGKLP] can be computed effectively, in nearly linear time.

The proof in [BGKLP] proceeds by the following steps.

(i) Let L be a lifting of the generators of the composition factors to G. Let M be a subset of L of size $O(\log |G|)$ which also generates G.

(ii) Let S be a generating set of G such that any element of G can be reached from S by a straight-line program of length $O(\log |G|)$. Write straight-line programs from S to M.

(iii) Write straight-line programs from S to $O(\log^C |G|)$ elements of G. These are elements whose membership in N_{i+1} was tested for some $1 \leq i \leq m-1$ in (i)–(iii) in the proof of Theorem 1.7. Now the presentation in [BGKLP] is obtained in $O(\log^{C+1} |G|)$ deterministic time.

In the proof of Theorem 1.7, we have seen that, in permutation groups with the composition factor restrictions of Corollary 1.8, presentations of length $O(\log^2 |N_i\varphi_i|)$ for the composition factors $N_i\varphi_i$ can be written by a nearly linear time algorithm. Hence, we can apply the result of [BGKLP] with the value $C = 2$. Moreover, the generating sets S_i^{**} of the composition factors constructed in the proof of Theorem 1.7 satisfy that $\bigcup_i S_i^{**}\varphi_i^{-1}$ has $O(\log |G|)$ elements, so we can choose $M := L$ in (i) above. We shall show that any given $g \in G$ can be reached from $\bigcup_i S_i^{**}\varphi_i^{-1}$ by a straight-line program of length $O(\log |G|)$, and such a straight-line program can be computed in nearly linear time. This means that we can choose $S := L = M$ in (ii), and a presentation of G of length $O(\log^3 |G|)$ can be written in nearly linear time, as indicated in (iii). Therefore, all that remains is to prove the following proposition.

Proposition 8.5 *Let $G \leq S_n$ be a permutation group, and suppose that the following have already been computed: a composition series $G = N_1 \vartriangleright N_2 \vartriangleright \cdots \vartriangleright N_m = 1$, homomorphisms $\varphi_i \colon N_i \to S_n$ with $\ker \varphi_i = N_{i+1}$, and presentations using generating sets $S_i^{**} \subset N_i\varphi_i$, as in the proof of Theorem 1.7. Then any $g \in G$ can be reached from $\bigcup_i S_i^{**}\varphi_i^{-1}$ by a straight-line program of length $O(\log |G|)$, and such a straight-line program can be computed in nearly linear time.*

Proof. By induction on $i = 1, 2, \ldots, m$, we construct a straight-line program of length $O(\log(|G|/|N_i|))$ to some $g_i \in G$ such that $gg_i^{-1} \in N_i$. Let $g_1 := 1$. If g_i has already been obtained for some i, then (as in the proof of Theorem 1.7) write a straight-line program of length $O(\log |N_i/N_{i+1}|)$ from S_i^{**} to $(gg_i^{-1})\varphi_i$. Mirror this straight-line program starting from $S_i^{**}\varphi_i^{-1}$, producing an element $h_i \in N_i$. Here, $gg_i^{-1}h_i^{-1} \in N_{i+1}$, and we can define $g_{i+1} := h_i g_i$. As we saw in the proof of Theorem 1.7, the straight-line program from S_i^{**} to $(gg_i^{-1})\varphi_i$ can be computed in nearly linear time; the rest of the algorithm clearly runs within this time bound. $\square$

9 Concluding remarks

9.1 Improvements?

There is a great deal of room for improvement in the algorithms presented here:

- We were not able to use L and Q directly without any recursive call. Yet, L is a classical group, and Q or Q/T is its natural module.

- The same work was done many times. Thus, we started with G, obtained a subgroup J, and then used it to construct a subgroup L. Within L another group J was found, and so on.

  Similarly, once we obtained a transvection or root element for G we obtained one (in fact, a whole transvection group or long root group) for L. Hence, this part of the recursive call can be sped up, but we still need a new "τ" for L.

  Less significantly, we constructed almost d copies of $\mathrm{GF}(q)$.

- Our probability estimates for generating classical groups were, in general, poor. It might be interesting to have better estimates (in fact, estimates with correct error terms), even just for generation by transvections. In any case, we certainly do not expect that there will be any need to test more than 2^{20} random elements, as suggested in **4.2.2**.

- We assumed that the characteristic p was part of the input. This can be replaced by other assumptions, e.g., that exact group element orders can be found [KS].

- Finally, the most important failing of our approach was that the timing was not quite polynomial in the input length: small powers of q remained in the timing estimates. However, entirely new ideas would be required to study whether there is a polynomial time algorithm for Theorem 1.1, at least if d is not too small.

Thus, no claim is made here that we have presented algorithms that are either practical or the final word in algorithms of this sort. Quite the contrary, there is no doubt in our minds that greater efficiency is possible from various points of view: practical, theoretical, and expository. As mentioned

159

in the introduction, the Monte Carlo algorithm in [CFL] leads to better timing estimates than ours when $G \cong \mathrm{PSL}(d, 2)$ for some (unknown) d, namely $O(\xi N^{1/2} + \mu N + N^{3/2})$. More recently, [BCFL] obtained a Las Vegas algorithm for the same question when $G \cong \mathrm{PSL}(d, q)$. These algorithms — perhaps combined with ideas from [CLG2] — need to be investigated further to see how to adapt the methodology in order to provide significant improvements to our timings.

One significant difference between our approach and that in [CFL] is our dependence on the recursive nature of the classical groups, which is entirely avoided in [CFL]. This dependence can be viewed as using a suitable maximal parabolic subgroup of G, its unipotent radical Q and the derived subgroup L of its Levi factor [Ca]. A much more Lie-theoretic point of view is presently being used in [KM] for exceptional groups of Lie type. (That paper uses a maximal torus of L in order to handle Q or Q/T. However, that method cannot be used for very small q.)

We note that alternating and symmetric groups can be recognized by a polynomial time Las Vegas algorithm ([BeB] and [BLNPS]).

Matrix groups.   Subgroups of $\mathrm{GL}(a, F)$ have a property not available in Theorem 1.1: If G is a a quasisimple subgroup of $\mathrm{GL}(a, F)$ then a basis of Q or $Q/\mathrm{Z}(Q)$ can be found quickly (this follows readily from results in [Lu]). When Q is abelian this observation has been noted as one of the two "gray box" assumptions in [CLG2] (the other assumption is that exact orders of elements can be found quickly). However, at present this has only led to a very minor improvement in timings in this paper. This deficiency will undergo further investigation.

9.2   Applications

- Corollary 1.6 has been used in [Mo1] to obtain Monte Carlo nearly linear algorithms that find and conjugate Sylow subgroups of permutation groups none of whose composition factors is an exceptional group of Lie type. Handling the latter groups in the context of Corollary 1.6, or more generally Theorem 1.1, will extend those Sylow algorithms to all permutation groups (cf. [KM]).

- There is a significant application of Theorem 1.1 within the matrix recognition project. Aschbacher [As1] classified the subgroups of general linear groups into several categories, and recent practical computations with matrix groups center around the problem of finding a category to which a given matrix group belongs [NP, NiP1, NiP2, CLG1, CLG2, HR1, HR2, HLOR1, HLOR2, LGO]. One of Aschbacher's categories is that the subgroup G of $GL(n, p^e)$ has a normal quasisimple classical matrix group, in its natural representation. In another category, $T \leq G/Z(G) \leq \mathrm{Aut}\,T$ for some simple group T, where the characteristic of T is not necessarily p. This case was not considered yet in the above mentioned references, and we expect that a practical version of Theorem 1.1 will be applied here. At this point versions of the algorithms in Sections 3 and 5 have been implemented in GAP.

Acknowlegdement: We are grateful to P. Brooksbank for many helpful comments concerning a variety of sections, and to R. A. Liebler for a very helpful suggestion leading to improvements in **6.6.1**.

References

[Ar]   E. Artin, The orders of the classical simple groups. Comm. Pure Appl. Math. 8 (1955) 455-472.

[As1]   M. Aschbacher, On the maximal subgroups of the finite classical groups. Invent. Math. 76 (1984) 469–514.

[As2]   M. Aschbacher, Finite group theory. Cambridge U. Press, Cambridge 1986.

[Ba1]   L. Babai, Local expansion of vertex-transitive graphs and random generation in finite groups, pp. 164–174 in: Proc. ACM Symp. on Theory of Computing 1991.

[Ba2]   L. Babai, Randomization in group algorithms: conceptual questions, pp. 1–17 in: Groups and Computation II, Proceedings of a DIMACS Workshop (eds. L. Finkelstein and W. M. Kantor), AMS 1997.

[BaB]   L. Babai and R. Beals, A polynomial-time theory of black-box groups I, pp. 30–64, in: Groups St Andrews 1997 in Bath, I (eds. C. M. Campbell, E. F. Robertson, N. Ruskuc and G. C. Smith), LMS Lecture Note Series 260, Cambridge U. Press 1999.

[BCFLS]   L. Babai, G. Cooperman, L. Finkelstein, E. M. Luks and Á Seress, Fast Monte Carlo algorithms for permutation groups. J. Comp. Syst. Sci. 50 (1995), 296–308.

[BCFS]   L. Babai, G. Cooperman, L. Finkelstein and Á Seress, Nearly linear time algorithms for permutation groups with a small base, pp. 200–209 in: Proc. Int. Symp. Symbolic and Algebraic Computation, ACM 1991.

[BGKLP]   L. Babai, A. J. Goodman, W. M. Kantor, E. M. Luks and P. P. Pálfy, Short presentations for finite groups. J. Algebra 194 (1997) 79–112.

[BKPS]   L. Babai, W. M. Kantor, P. Pálfy and Á. Seress, Black-box recognition of finite simple groups of Lie type by statistics of element orders (in preparation).

162

[BLS1]   L. Babai, E. M. Luks and Á. Seress, Permutation groups in NC, pp. 409–420 in: Proc. ACM Symp. on Theory of Computing 1987.

[BLS2]   L. Babai, E. M. Luks and Á. Seress, Fast management of permutation groups I. SIAM J. Computing 26 (1997) 1310–1342.

[BLS3]   L. Babai, E. M. Luks and Á. Seress, Fast management of permutation groups II (in preparation).

[BSz]   L. Babai and E. Szemerédi, On the complexity of matrix group problems, I, pp. 229–240 in: Proc. IEEE Symp. Found. Comp. Sci. 1984.

[Be]   R. Beals, Towards polynomial time algorithms for matrix groups, pp. 31–54 in: Groups and Computation II, Proceedings of a DIMACS Workshop (eds. L. Finkelstein and W. M. Kantor), AMS 1997.

[BeB]   R. Beals and L. Babai, Las Vegas algorithms for matrix groups, pp. 427–436 in: Proc. IEEE Symp. Found. Comp. Sci. 1993.

[BLNPS]   R. Beals, C. R. Leedham-Green, A. C. Niemeyer, C. E. Praeger, and Á. Seress, A mélange of black box algorithms for recognising finite symmetric and alternating groups (in preparation).

[BeS]   R. Beals and Á. Seress, Structure forest and composition factors for small base groups in nearly linear time, pp. 116–125 in: Proc. ACM Symp. on Theory of Computing 1992.

[BCP]   W. Bosma, J. Cannon and C. Playoust, The Magma algebra system I: The user language. J. Symb. Comp. 24 (1997) 235–265.

[BCFL]   S. Bratus, G. Cooperman, L. Finkelstein and S. Linton (in preparation).

[Carm]   R. D. Carmichael, Abstract definitions of the symmetric and alternating groups and certain other permutation groups. Quart. J. Math. 49 (1923) 226–270.

[Ca]   R. Carter, Simple groups of Lie type. Wiley, London–New York–Sydney–Toronto 1972.

[Ce]   F. Celler, Matrixgruppenalgorithmen in GAP. Ph. D. thesis, RWTH Aachen 1997.

[CLG1]   F. Celler and C. R. Leedham-Green, A non-constructive recognition algorithm for the special linear and other classical groups, pp. 61–67 in: Groups and Computation II, Proceedings of a DIMACS Workshop (eds. L. Finkelstein and W. M. Kantor), AMS 1997.

[CLG2]   F. Celler and C. R. Leedham-Green, A constructive recognition algorithm for the special linear group, pp. 11–26 in: The atlas of finite groups: ten years on (Birmingham, 1995), LMS Lecture Note Series 249, Cambridge U. Press 1998.

[CLMNO]   F. Celler, C. R. Leedham-Green, S. H. Murray, A. C. Niemeyer and E. A. O'Brien, Generating random elements of a finite group. Comm. in Alg. 23 (1995) 4931–4948.

[Che]   H. Chernoff, A measure of asymptotic efficiency for tests of a hypothesis based on the sum of observations. Ann. Math. Statistics 23 (1952) 493–507.

[Con]   J. H. Conway, A tabulation of some information concerning finite fields, pp. 37–50 in: Computers in Mathematical Research (eds. R. F. Churchhouse and J. C. Herz), North Holland 1968.

[CCNPW]   J. H. Conway, R. T. Curtis, S. P. Norton, R. A. Parker and R. A. Wilson, Atlas of finite groups. Clarendon Press, Oxford 1985.

[CFL]   G. Cooperman, L. Finkelstein and S. Linton, Recognizing $GL_n(2)$ in non-standard representation, pp. 85–100 in: Groups and Computation II, Proceedings of a DIMACS Workshop (eds. L. Finkelstein and W. M. Kantor), AMS 1997.

[Co]   B. N. Cooperstein, Minimal degree for a permutation representation of a classical group. Israel J. Math. 30 (1978) 213–235.

[Cu]   C. W. Curtis, Central extensions of groups of Lie type. J. reine und angew. Math. 220 (1965) 174–185.

[Di]   L. E. Dickson, Linear groups, with an exposition of Galois theory. Teubner, Leipzig 1901; reprint Dover, New York 1958.

[GAP99] The GAP Group, *GAP – Groups, Algorithms, and Programming, Version 4.1*, Aachen/St Andrews 1999 (URL: http://www-gap.dcs.st-and.ac.uk/~gap).

[Gr] R. L. Griess, Schur multipliers of finite simple groups of Lie type. Trans. AMS 183 (1973) 355–421.

[Ha] R. W. Hartley, Determination of the ternary linear collineation groups whose coefficients lie in $GF(2^n)$. Annals of Math. 27 (1926) 140–158.

[HLOR1] D. F. Holt, C. R. Leedham-Green, E. A. O'Brien and S. Rees, Testing matrix groups for primitivity. J. Algebra 184 (1996) 795–817.

[HLOR2] D. F. Holt, C. R. Leedham-Green, E. A. O'Brien and S. Rees, Computing matrix group decompositions with respect to a normal subgroup. J. Algebra 184 (1996) 818–838.

[HR1] D. F. Holt and S. Rees, An implementation of the Neumann–Praeger algorithm for the recognition of special linear groups. J. Experimental Math. 1 (1992) 237–242.

[HR2] D. F. Holt and S. Rees, Testing modules for irreducibility. J. Austral. Math. Soc. (Ser. A) 57 (1994) 1–16.

[Ka1] W. M. Kantor, Permutation representations of the finite classical groups of small degree or rank. J. Algebra 60 (1979) 158–168.

[Ka2] W. M. Kantor, Subgroups of classical groups generated by long root elements. Trans. AMS 248 (1979) 347–379.

[Ka3] W. M. Kantor, Sylow's theorem in polynomial time. J. Comp. Syst. Sci. 30 (1985) 359–394.

[Ka4] W. M. Kantor, Geometry in computer algebra systems (unpublished manuscript for Magma Conf., London 1993; URL: http://www.uoregon.edu/~kantor/PAPERS/magmapaper.ps).

[Ka5] W. M. Kantor, Simple groups in computational group theory, pp. 77–86 in: Proc. International Congress of Mathematicans, Berlin 1998, Vol. II.

[KaLu]   W. M. Kantor and A. Lubotzky, The probability of generating a finite classical group. Geom. Ded. 36 (1990) 67-87.

[KLM]   W. M. Kantor, E. M. Luks and P. D. Mark, Sylow subgroups in parallel. J. Algorithms 31 (1999) 132–195.

[KM]   W. M. Kantor and K. Magaard, Black box exceptional groups of Lie type (in preparation).

[KP]   W. M. Kantor and T. Penttila, Reconstructing simple group actions, pp. 147–180 in: Proc. Conf. "Geometric Group Theory Down Under" (eds. J. Cossey et al.), de Gruyter 1999.

[KS]   W. M. Kantor and Á. Seress, Prime power graphs for groups of Lie type (in preparation).

[KL]   P. B. Kleidman and M. W. Liebeck, The subgroup structure of the finite classical groups. LMS Lecture Note Series 129, Cambridge U. Press 1990.

[LG]   C. R. Leedham-Green, Talk and rough draft at Oberwolfach Meeting on Computational Group Theory, 1992.

[LGO]   C. R. Leedham-Green and E. A. O'Brien, Recognising tensor products of matrix groups. Int. J. Algebra Comput. 7 (1997) 541–559.

[LN]   R. Lidl and H. Niederreiter, Finite Fields. Encyclopedia of Math. and its Applications, vol. 20, Addison–Wesley 1983.

[Lu]   E. M. Luks, Computing in solvable matrix groups, pp. 111–120. in: Proc. IEEE Symp. Found. Comp. Sci. 1992.

[Ma]   P. D. Mark, Sylow's theorem and parallel computation. Ph. D. thesis, U. of Oregon 1993.

[Mc1]   J. McLaughlin, Some groups generated by transvections. Arch. Math. 18 (1967) 364–368.

[Mc2]   J. McLaughlin, Some subgroups of $SL_n(F_2)$. Illinois J. Math. 13 (1969) 108–115.

[Mi1]   H. H. Mitchell, Determination of the ordinary and modular ternary linear groups. Trans. AMS 12 (1911) 207–242.

[Mi2]   H.H. Mitchell, The subgroups of the quaternary abelian linear group, Trans. AMS 15 (1914) 379–396.

[Mo]   E. H. Moore, The subgroups of the generalized modular group. Decennial Publ. U. Chicago 9 (1904) 141–190.

[Mo1]   P. Morje, A nearly linear algorithm for Sylow subgroups of permutation groups. Ph.D. thesis, The Ohio State U. 1995.

[Mo2]   P. Morje, A nearly linear algorithm for Sylow subgroups of permutation groups, pp. 270–277 in: Proc. Int. Symp. Symbolic and Algebraic Computation, ACM 1995.

[Mo3]   P. Morje, On nearly linear time algorithms for Sylow subgroups of small-base permutation groups, pp. 257–272 in: Groups and Computation II, Proceedings of a DIMACS Workshop (eds. L. Finkelstein and W. M. Kantor), AMS 1997.

[NP]   P. M. Neumann and C. E. Praeger, A recognition algorithm for special linear groups. Proc. London Math. Soc. 65 (1992), 555–603.

[NiP1]   A. C. Niemeyer and C. E. Praeger, Implementing a recognition algorithm for classical groups, pp. 273–296 in: Groups and Computation II, Proceedings of a DIMACS Workshop (eds. L. Finkelstein and W. M. Kantor), AMS 1997.

[NiP2]   A. C. Niemeyer and C. E. Praeger, A recognition algorithm for classical groups over finite fields. Proc. London Math. Soc. 77 (1998) 117–169.

[Ser1]   Á. Seress, Nearly linear time algorithms for permutation groups: an interplay between theory and practice. Acta Appl. Math. 52 (1998), 183–207.

[Ser2]   Á. Seress, Permutation Group Algorithms. Cambridge U. Press, to appear.

[Si1]   C. C. Sims, Computational methods in the study of permutation groups, pp. 169–183 in: Computational problems in abstract algebra (ed. J. Leech), Pergamon Press 1970.

[Si2]   C. C. Sims, Computation with permutation groups, pp. 23–28 in: Proc. 2nd Symp. on Symb. and Alg. Manipulation (ed. S. R. Petrick), ACM 1971.

[St]   R. Steinberg, Generators, relations and coverings of algebraic groups, II. J. Algebra 71 (1981) 527–543.

[Ta]   D. E. Taylor, The geometry of the classical groups. Heldermann, Berlin 1992.

[Wi]   A. Wiman, Bestimmung aller Untergruppen einer doppelt unendlichen Reihe von einfachen Gruppen. Bihang till K. Svenska Vet.-Akad. Handl. 25 (1899) 1–47.

[Zs]   K. Zsigmondy, Zur Theorie der Potenzreste. Monatsh. Math. Phys. 3 (1892) 265–284.

University of Oregon, Eugene, OR 97403
    kantor@math.uoregon.edu

The Ohio State University, Columbus, OH 43210
    akos@math.ohio-state.edu

Editorial Information

To be published in the *Memoirs*, a paper must be correct, new, nontrivial, and significant. Further, it must be well written and of interest to a substantial number of mathematicians. Piecemeal results, such as an inconclusive step toward an unproved major theorem or a minor variation on a known result, are in general not acceptable for publication. Papers appearing in *Memoirs* are generally longer than those appearing in *Transactions*, which shares the same editorial committee.

As of September 30, 2000, the backlog for this journal was approximately 11 volumes. This estimate is the result of dividing the number of manuscripts for this journal in the Providence office that have not yet gone to the printer on the above date by the average number of monographs per volume over the previous twelve months, reduced by the number of volumes published in four months (the time necessary for preparing a volume for the printer). (There are 6 volumes per year, each containing at least 4 numbers.)

A Consent to Publish and Copyright Agreement is required before a paper will be published in the *Memoirs*. After a paper is accepted for publication, the Providence office will send a Consent to Publish and Copyright Agreement to all authors of the paper. By submitting a paper to the *Memoirs*, authors certify that the results have not been submitted to nor are they under consideration for publication by another journal, conference proceedings, or similar publication.

Information for Authors

Memoirs are printed from camera copy fully prepared by the author. This means that the finished book will look exactly like the copy submitted.

The paper must contain a *descriptive title* and an *abstract* that summarizes the article in language suitable for workers in the general field (algebra, analysis, etc.). The *descriptive title* should be short, but informative; useless or vague phrases such as "some remarks about" or "concerning" should be avoided. The *abstract* should be at least one complete sentence, and at most 300 words. Included with the footnotes to the paper should be the 2000 *Mathematics Subject Classification* representing the primary and secondary subjects of the article. The classifications are accessible from www.ams.org/msc/. The list of classifications is also available in print starting with the 1999 annual index of *Mathematical Reviews*. The Mathematics Subject Classification footnote may be followed by a list of *key words and phrases* describing the subject matter of the article and taken from it. Journal abbreviations used in bibliographies are listed in the latest *Mathematical Reviews* annual index. The series abbreviations are also accessible from www.ams.org/publications/. To help in preparing and verifying references, the AMS offers MR Lookup, a Reference Tool for Linking, at www.ams.org/mrlookup/. When the manuscript is submitted, authors should supply the editor with electronic addresses if available. These will be printed after the postal address at the end of the article.

Electronically prepared manuscripts. The AMS encourages electronically prepared manuscripts, with a strong preference for $\mathcal{A}_{\mathcal{M}}\mathcal{S}$-LaTeX. To this end, the Society has prepared $\mathcal{A}_{\mathcal{M}}\mathcal{S}$-LaTeX author packages for each AMS publication. Author packages include instructions for preparing electronic manuscripts, the *AMS Author Handbook*, samples, and a style file that generates the particular design specifications of that publication series. Though $\mathcal{A}_{\mathcal{M}}\mathcal{S}$-LaTeX is the highly preferred format of TeX, author packages are also available in $\mathcal{A}_{\mathcal{M}}\mathcal{S}$-TeX.

Authors may retrieve an author package from e-MATH starting from `www.ams.org/tex/` or via FTP to `ftp.ams.org` (login as `anonymous`, enter username as password, and type `cd pub/author-info`). The *AMS Author Handbook* and the *Instruction Manual* are available in PDF format following the author packages link from `www.ams.org/tex/`. The author package can be obtained free of charge by sending email to `pub@ams.org` (Internet) or from the Publication Division, American Mathematical Society, P.O. Box 6248, Providence, RI 02940-6248. When requesting an author package, please specify $\mathcal{A}_{\mathcal{M}}\mathcal{S}$-LaTeX or $\mathcal{A}_{\mathcal{M}}\mathcal{S}$-TeX, Macintosh or IBM (3.5) format, and the publication in which your paper will appear. Please be sure to include your complete mailing address.

Sending electronic files. After acceptance, the source file(s) should be sent to the Providence office (this includes any TeX source file, any graphics files, and the DVI or PostScript file).

Before sending the source file, be sure you have proofread your paper carefully. The files you send must be the EXACT files used to generate the proof copy that was accepted for publication. For all publications, authors are required to send a printed copy of their paper, which exactly matches the copy approved for publication, along with any graphics that will appear in the paper.

TeX files may be submitted by email, FTP, or on diskette. The DVI file(s) and PostScript files should be submitted only by FTP or on diskette unless they are encoded properly to submit through email. (DVI files are binary and PostScript files tend to be very large.)

Electronically prepared manuscripts can be sent via email to `pub-submit@ams.org` (Internet). The subject line of the message should include the publication code to identify it as a Memoir. TeX source files, DVI files, and PostScript files can be transferred over the Internet by FTP to the Internet node `e-math.ams.org` (130.44.1.100).

Electronic graphics. Comprehensive instructions on preparing graphics are available at `www.ams.org/jourhtml/graphics.html`. A few of the major requirements are given here.

Submit files for graphics as EPS (Encapsulated PostScript) files. This includes graphics originated via a graphics application as well as scanned photographs or other computer-generated images. If this is not possible, TIFF files are acceptable as long as they can be opened in Adobe Photoshop or Illustrator. No matter what method was used to produce the graphic, it is necessary to provide a paper copy to the AMS.

Authors using graphics packages for the creation of electronic art should also avoid the use of any lines thinner than 0.5 points in width. Many graphics packages allow the user to specify a "hairline" for a very thin line. Hairlines often look acceptable when proofed on a typical laser printer. However, when produced on a high-resolution laser imagesetter, hairlines become nearly invisible and will be lost entirely in the final printing process.

Screens should be set to values between 15% and 85%. Screens which fall outside of this range are too light or too dark to print correctly. Variations of screens within a graphic should be no less than 10%.

Inquiries. Any inquiries concerning a paper that has been accepted for publication should be sent directly to the Electronic Prepress Department, American Mathematical Society, P. O. Box 6248, Providence, RI 02940-6248.

Selected Titles in This Series

For a complete list of titles in this series, visit the
AMS Bookstore at **www.ams.org/bookstore/**.